BOTANY
MULTIPLE CHOICE QUESTIONS AND ANSWERS

For
HIGH SCHOOLS AND COLLEGES

An indispensable guide for maximum blasting!

Prepared and compiled by;

Professor Carl William Matthias

Contents

ACKNOWLEDGEMENT

We would like to express our sincere gratitude to all the individuals and resources that have contributed to the creation of this multiple choice questions and answers book on botany for high schools and colleges. This comprehensive resource would not have been possible without their valuable support, expertise, and dedication.

We would also like to acknowledge the diligent efforts of the researchers, authors, and editors who meticulously crafted each question and answer, ensuring the clarity, coherence, and educational value of the material. Their commitment to excellence has greatly enriched this book and will undoubtedly benefit students at various academic levels.

Additionally, we are indebted to the educators, students, and reviewers who took the time to provide feedback and suggestions during the review process. Their input has played a crucial role in refining the content and enhancing its educational effectiveness.

It is with immense pride and gratitude that we present this multiple choice questions and answers book on botany, knowing that it will serve as a valuable resource for students, educators, and botany enthusiasts alike. May it contribute to a deeper understanding and appreciation of the field of plants.

Thank you all for your unwavering support, dedication, and passion.

CHAPTER 1: PLANT ANATOMY AND MORPHOLOGY

PLANT CELLS

1. Which organelle is responsible for converting light energy into chemical energy in plant cells? a) Nucleus b) Mitochondria c) Golgi apparatus d) Chloroplast

2. Which of the following structures is not found in a plant cell? a) Cell wall b) Vacuole c) Nucleus d) Centriole

3. Which plant cell organelle is responsible for the synthesis of proteins? a) Ribosome b) Nucleus c) Golgi apparatus d) Lysosome

4. What is the function of the cell wall in plant cells? a) To provide support and protection b) To store water and nutrients c) To produce energy d) To control cell division

5. Which of the following is true about plant cell vacuoles? a) They are responsible for photosynthesis. b) They contain genetic material. c) They store water, ions, and nutrients. d) They are involved in protein synthesis.

6. In which part of the plant cell does photosynthesis primarily occur? a) Nucleus b) Mitochondria c) Chloroplast d) Golgi apparatus

7. What is the function of the Golgi apparatus in plant cells? a) To produce energy b) To transport proteins c) To store water d) To control cell division

8. Which of the following is not a function of the mitochondria in plant cells? a) Energy production b) Cellular respiration c) Protein synthesis d) ATP synthesis

9. What is the primary role of the nucleus in plant cells? a) To store water and nutrients b) To control cell division c) To produce energy d) To synthesize proteins and store genetic material

10. Which organelle is responsible for breaking down waste materials and cellular debris in plant cells? a) Ribosome b) Nucleus c) Lysosome d) Chloroplast

11. What is the function of the endoplasmic reticulum in plant cells? a) To produce energy b) To transport proteins c) To store water d) To control cell division

12. Which of the following is not a characteristic of plant cells? a) Cell wall b) Chloroplasts c) Nucleus d) Centrioles

13. Which organelle is responsible for the synthesis of lipids in plant cells? a) Ribosome b) Nucleus c) Lysosome d) Smooth endoplasmic reticulum

14. Which of the following is true about plant cell mitochondria? a) They contain genetic material. b) They are involved in photosynthesis. c) They store water and ions. d) They produce energy through cellular respiration.

15. Which organelle is responsible for packaging and sorting proteins in plant cells? a) Ribosome b) Nucleus c) Golgi apparatus d) Rough endoplasmic reticulum

16. What is the function of the peroxisomes in plant cells? a) To produce energy b) To detoxify harmful substances c) To store water d) To control cell division

17. Which of the following is not a function of plant cell chloroplasts? a) Energy production b) Photosynthesis c) Cellular respiration d) Synthesis of sugars

18. Where are plant cell ribosomes primarily located? a) Nucleus b) Mitochondria c) Chloroplast d) Rough endoplasmic reticulum

19. What is the function of the cytoskeleton in plant cells? a) To provide structural support and maintain cell shape b) To store water and nutrients c) To produce energy d) To control cell division

20. Which of the following is not a component of the plant cell cytoskeleton? a) Microtubules b) Microfilaments c) Intermediate filaments d) Centrioles

21. What is the function of the plasmodesmata in plant cells? a) To transport water and nutrients between cells b) To produce energy c) To store water and ions d) To control cell division

22. Which organelle is responsible for the synthesis of cellulose in plant cells? a) Ribosome b) Nucleus c) Lysosome d) Golgi apparatus

23. What is the function of the amyloplasts in plant cells? a) To produce energy b) To store starch c) To detoxify harmful substances d) To control cell division

24. Which of the following is not a function of plant cell vacuoles? a) Storage of water and ions b) Detoxification of harmful substances c) Maintenance of turgor pressure d) Breakdown of macromolecules

25. What is the primary role of the nucleolus in plant cells? a) To store water and nutrients b) To synthesize proteins c) To produce energy d) To assemble ribosomes

26. Which organelle is responsible for lipid metabolism and calcium storage in plant cells? a) Ribosome b) Nucleus c) Peroxisome d) Smooth endoplasmic reticulum

27. What is the function of the plasmalemma in plant cells? a) To protect the cell from external damage b) To produce energy c) To store water and ions d) To control cell division

28. Which of the following is not a component of the plant cell wall? a) Cellulose b) Lignin c) Starch d) Pectin

29. What is the function of the nucleoplasm in plant cells? a) To produce energy b) To store water and nutrients c) To protect and support the nucleus d) To control cell division

30. Which organelle is responsible for the breakdown of macromolecules and cellular recycling in plant cells? a) Ribosome b) Nucleus c) Lysosome d) Chloroplast

31. Which organelle in plant cells is responsible for the production of ATP through cellular respiration? a) Ribosome b) Nucleus c) Chloroplast d) Mitochondria

32. What is the function of the plasmids in plant cells? a) To store water and nutrients b) To synthesize proteins c) To control cell division d) To store extra DNA molecules

33. Which of the following is responsible for the movement of water and nutrients between plant cells? a) Plasmodesmata b) Centrioles c) Golgi apparatus d) Smooth endoplasmic reticulum

34. What is the primary role of the peroxisomes in plant cells? a) To produce energy b) To break down fatty acids c) To store water and ions d) To synthesize proteins

35. Which organelle in plant cells is responsible for the formation of spindle fibers during cell division? a) Ribosome b) Nucleus c) Centriole d) Lysosome

36. What is the function of the plasmalemma in plant cells? a) To transport water and nutrients b) To produce energy c) To store water and ions d) To protect the cell from external damage

37. Which of the following is responsible for the synthesis of cell wall components in plant cells? a) Ribosome b) Nucleus c) Golgi apparatus d) Smooth endoplasmic reticulum

38. What is the primary function of the leucoplasts in plant cells? a) To produce energy b) To store water and ions c) To synthesize lipids and oils d) To control cell division

39. Which organelle in plant cells is responsible for the breakdown of hydrogen peroxide? a) Ribosome b) Nucleus c) Peroxisome d) Chloroplast

40. What is the function of the tonoplast in plant cells? a) To transport water and solutes b) To produce energy c) To store water and ions d) To control cell division

41. Which of the following is responsible for the movement of organelles and vesicles within plant cells? a) Plasmodesmata b) Microtubules c) Golgi apparatus d) Smooth endoplasmic reticulum

42. What is the primary role of the dictyosomes in plant cells? a) To produce energy b) To synthesize lipids and oils c) To store water and ions d) To modify and package proteins

43. Which organelle in plant cells is responsible for the synthesis of ribosomal RNA? a) Ribosome b) Nucleus c) Lysosome d) Nucleolus

44. What is the function of the contractile vacuole in plant cells? a) To produce energy b) To store water and nutrients c) To control cell division d) To regulate water balance

45. Which of the following is responsible for the production of lignin in plant cells? a) Ribosome b) Nucleus c) Golgi apparatus d) Smooth endoplasmic reticulum

46. What is the primary function of the glyoxysomes in plant cells? a) To produce energy b) To break down stored fats c) To store water and ions d) To synthesize proteins

47. Which organelle in plant cells is responsible for the breakdown of complex carbohydrates? a) Ribosome b) Nucleus c) Lysosome d) Vacuole

48. What is the function of the phragmoplast in plant cells? a) To produce energy b) To facilitate cell division c) To store water and ions d) To synthesize lipids and oils

49. Which of the following is responsible for the synthesis of nucleotides in plant cells? a) Ribosome b) Nucleus c) Golgi apparatus d) Smooth endoplasmic reticulum

50. What is the primary function of the hydrogenosomes in plant cells? a) To produce energy b) To break down complex carbohydrates c) To store water and ions d) To control cell division

ANSWERS

1. d) Chloroplast
2. d) Centriole
3. a) Ribosome
4. a) To provide support and protection
5. c) They store water, ions, and nutrients.
6. c) Chloroplast
7. b) To transport proteins
8. c) Protein synthesis
9. d) To synthesize proteins and store genetic material
10. c) Lysosome
11. b) To transport proteins
12. d) Centrioles
13. d) Smooth endoplasmic reticulum
14. d) They produce energy through cellular respiration.
15. c) Golgi apparatus
16. b) To detoxify harmful substances
17. c) Cellular respiration
18. d) Rough endoplasmic reticulum
19. a) To provide structural support and maintain cell shape
20. d) Centrioles
21. a) To transport water and nutrients between cells
22. d) Golgi apparatus
23. b) To store starch
24. b) Detoxification of harmful substances
25. d) To assemble ribosomes

26. c) Peroxisome

27. a) To protect the cell from external damage

28. c) Starch

29. c) To protect and support the nucleus

30. c) Lysosome

31. d) Mitochondria

32. d) To store extra DNA molecules

33. a) Plasmodesmata

34. b) To break down fatty acids

35. c) Centriole

36. a) To transport water and nutrients

37. c) Golgi apparatus

38. c) To synthesize lipids and oils

39. c) Peroxisome

40. c) To store water and ions

41. b) Microtubules

42. d) To modify and package proteins

43. d) Nucleolus

44. d) To regulate water balance

45. c) Golgi apparatus

46. b) To break down stored fats

47. c) Lysosome

48. b) To facilitate cell division

49. d) Smooth endoplasmic reticulum

50. a) To produce energy

PLANT TISSUES

1. Which of the following is the primary function of plant tissues? a) Support and protection b) Reproduction c) Photosynthesis d) Water absorption

2. What is the main role of meristematic tissues in plants? a) Support and conduction b) Photosynthesis c) Growth and cell division d) Water storage

3. Which type of tissue is responsible for transporting water and nutrients throughout the plant? a) Epidermal tissue b) Ground tissue c) Vascular tissue d) Meristematic tissue

4. What is the function of the epidermal tissue in plants? a) Photosynthesis b) Storage of water and nutrients c) Protection and prevention of water loss d) Cell division and growth

5. Which type of tissue forms the outermost layer of a plant and provides a protective covering? a) Meristematic tissue b) Epidermal tissue c) Vascular tissue d) Ground tissue

6. What is the function of the vascular tissue in plants? a) Photosynthesis b) Support and conduction of water and nutrients c) Protection and prevention of water loss d) Cell division and growth

7. Which type of tissue is responsible for the synthesis and storage of food in plants? a) Meristematic tissue b) Epidermal tissue c) Vascular tissue d) Ground tissue

8. What is the main role of collenchyma tissue in plants? a) Support and flexibility b) Photosynthesis c) Water absorption d) Storage of nutrients

9. Which type of tissue is responsible for providing mechanical support to young and growing parts of a plant? a) Meristematic tissue b) Epidermal tissue c) Vascular tissue d) Ground tissue

10. What is the function of the cork cambium in plants? a) Production of cork cells for protection b) Photosynthesis c) Water absorption d) Storage of nutrients

11. Which type of tissue is responsible for the regeneration of damaged plant parts? a) Meristematic tissue b) Epidermal tissue c) Vascular tissue d) Ground tissue

12. What is the main function of the ground tissue in plants? a) Support and conduction b) Photosynthesis c) Water absorption d) Storage of nutrients

13. Which type of tissue forms the pith and cortex regions in plant stems and roots? a) Meristematic tissue b) Epidermal tissue c) Vascular tissue d) Ground tissue

14. What is the function of the companion cells in plants? a) Support and conduction b) Photosynthesis c) Food storage and transport d) Water absorption

15. Which type of tissue is responsible for the production of secondary growth in woody plants? a) Meristematic tissue b) Epidermal tissue c) Vascular tissue d) Ground tissue

16. What is the main role of the periderm tissue in plants? a) Photosynthesis b) Water absorption c) Protection and prevention of water loss d) Cell division and growth

17. Which type of tissue is responsible for the formation of bark in plant stems and roots? a) Meristematic tissue b) Epidermal tissue c) Vascular tissue d) Cork cambium

18. What is the function of the sieve tubes in plants? a) Support and conduction b) Photosynthesis c) Food transport d) Water absorption

19. Which type of tissue is responsible for the production of xylem and phloem in plants? a) Meristematic tissue b) Epidermal tissue c) Vascular cambium d) Ground tissue

20. What is the main role of the endodermis tissue in plants? a) Photosynthesis b) Water absorption c) Protection and prevention of water loss d) Conduction of nutrients

21. Which type of tissue is responsible for the regulation of water movement in plant roots? a) Meristematic tissue b) Epidermal tissue c) Vascular tissue d) Endodermis

22. What is the function of the tracheids and vessel elements in plants? a) Support and conduction b) Photosynthesis c) Food storage and transport d) Water absorption

23. Which type of tissue is responsible for the formation of wood in plants? a) Meristematic tissue b) Epidermal tissue c) Vascular cambium d) Ground tissue

24. What is the main role of the cork cells in plants? a) Photosynthesis b) Water absorption c) Protection and prevention of water loss d) Cell division and growth

25. Which type of tissue is responsible for the formation of the pericycle and vascular cambium in plants? a) Meristematic tissue b) Epidermal tissue c) Vascular tissue d) Ground tissue

26. What is the function of the phloem tissue in plants? a) Support and conduction b) Photosynthesis c) Food transport d) Water absorption

27. Which type of tissue is responsible for the storage of starch, oils, and proteins in plants? a) Meristematic tissue b) Epidermal tissue c) Parenchyma tissue d) Ground tissue

28. What is the main role of the apical meristem in plants? a) Support and conduction b) Photosynthesis c) Growth in length d) Water absorption

29. Which type of tissue is responsible for the formation of new leaves and stems in plants? a) Meristematic tissue b) Epidermal tissue c) Vascular tissue d) Ground tissue

30. What is the function of the parenchyma tissue in plants? a) Support and conduction b) Photosynthesis c) Food storage and transport d) Water absorption

31. Which type of tissue is responsible for the storage of water and nutrients in plants? a) Meristematic tissue b) Epidermal tissue c) Vascular tissue d) Ground tissue

32. What is the main role of the cambium tissue in plants? a) Support and conduction b) Photosynthesis c) Growth in girth d) Water absorption

33. Which type of tissue is responsible for the production of cork cells in plants? a) Meristematic tissue b) Epidermal tissue c) Vascular tissue d) Cork cambium

34. In which type of tissue would you find casparian strips? a) Meristematic tissue b) Epidermal tissue c) Vascular tissue d) Ground tissue

35. Which type of tissue is responsible for the production of bark in plant stems and roots? a) Meristematic tissue b) Epidermal tissue c) Vascular tissue d) Cork cambium

36. What is the function of the sieve tubes in plants? a) Support and conduction b) Photosynthesis c) Food transport d) Water absorption

37. Which type of tissue is responsible for the production of xylem and phloem in plants? a) Meristematic tissue b) Epidermal tissue c) Vascular cambium d) Ground tissue

38. What is the main role of the endodermis tissue in plants? a) Photosynthesis b) Water absorption c) Protection and prevention of water loss d) Conduction of nutrients

39. Which type of tissue is responsible for the regulation of water movement in plant roots? a) Meristematic tissue b) Epidermal tissue c) Vascular tissue d) Endodermis

40. What is the function of the tracheids and vessel elements in plants? a) Support and conduction b) Photosynthesis c) Food storage and transport d) Water absorption

41. Which type of tissue is responsible for the formation of wood in plants? a) Meristematic tissue b) Epidermal tissue c) Vascular cambium d) Ground tissue

42. What is the main role of the cork cells in plants? a) Photosynthesis b) Water absorption c) Protection and prevention of water loss d) Cell division and growth

43. Which type of tissue is responsible for the formation of the pericycle and vascular cambium in plants? a) Meristematic tissue b) Epidermal tissue c) Vascular tissue d) Ground tissue

44. What is the function of the phloem tissue in plants? a) Support and conduction b) Photosynthesis c) Food transport d) Water absorption

45. Which type of tissue is responsible for the storage of starch, oils, and proteins in plants? a) Meristematic tissue b) Epidermal tissue c) Parenchyma tissue d) Ground tissue

46. What is the main role of the apical meristem in plants? a) Support and conduction b) Photosynthesis c) Growth in length d) Water absorption

47. Which type of tissue is responsible for the formation of new leaves and stems in plants? a) Meristematic tissue b) Epidermal tissue c) Vascular tissue d) Ground tissue

48. What is the function of the parenchyma tissue in plants? a) Support and conduction b) Photosynthesis c) Food storage and transport d) Water absorption

49. Which type of tissue is responsible for the storage of water and nutrients in plants? a) Meristematic tissue b) Epidermal tissue c) Vascular tissue d) Ground tissue

50. What is the main role of the cambium tissue in plants? a) Support and conduction b) Photosynthesis c) Growth in girth d) Water absorption

51. Which type of tissue is responsible for the production of cork cells in plants? a) Meristematic tissue b) Epidermal tissue c) Vascular tissue d) Cork cambium

52. What is the function of the endosperm tissue in plants? a) Support and conduction b) Photosynthesis c) Food storage and nourishment d) Water absorption

53. Which type of tissue is responsible for the formation of the seed coat in plants? a) Meristematic tissue b) Epidermal tissue c) Vascular tissue d) Ground tissue

54. What is the main role of the vascular cambium in plants? a) Support and conduction b) Photosynthesis c) Secondary growth d) Water absorption

55. Which type of tissue is responsible for the production of pollen grains in plants? a) Meristematic tissue b) Epidermal tissue c) Vascular tissue d) Ground tissue

56. What is the function of the glandular tissue in plants? a) Secretion of nectar and other substances b) Photosynthesis c) Storage of water and nutrients d) Water absorption

57. Which type of tissue is responsible for the production of flowers in plants? a) Meristematic tissue b) Epidermal tissue c) Vascular tissue d) Ground tissue

58. What is the main role of the pith tissue in plants? a) Support and conduction b) Photosynthesis c) Water absorption d) Storage of water and nutrients

59. Which type of tissue is responsible for the production of fruit in plants? a) Meristematic tissue b) Epidermal tissue c) Vascular tissue d) Ground tissue

60. What is the function of the pericycle tissue in plants? a) Support and conduction b) Photosynthesis c) Protection of the root d) Water absorption

61. Which type of tissue is responsible for the production of latex in plants? a) Meristematic tissue b) Epidermal tissue c) Vascular tissue d) Laticiferous tissue

62. What is the main role of the laticiferous tissue in plants? a) Support and conduction b) Photosynthesis c) Storage of water and nutrients d) Latex production

63. Which type of tissue is responsible for the production of fibers in plants? a) Meristematic tissue b) Epidermal tissue c) Vascular tissue d) Sclerenchyma tissue

64. What is the function of the sclerenchyma tissue in plants? a) Support and conduction b) Photosynthesis c) Storage of water and nutrients d) Mechanical strength and rigidity

65. Which type of tissue is responsible for the production of spines and thorns in plants? a) Meristematic tissue b) Epidermal tissue c) Vascular tissue d) Sclerenchyma tissue

66. What is the main role of the companion cells in plants? a) Support and conduction b) Photosynthesis c) Storage of water and nutrients d) Assistance in phloem transport

67. Which type of tissue is responsible for the production of cork in plants? a) Meristematic tissue b) Epidermal tissue c) Vascular tissue d) Cork cambium

68. What is the function of the epidermal tissue in plants? a) Support and conduction b) Photosynthesis c) Protection and prevention of water loss d) Storage of nutrients

69. Which type of tissue is responsible for the formation of the cuticle in plants? a) Meristematic tissue b) Epidermal tissue c) Vascular tissue d) Ground tissue

70. What is the main role of the mesophyll tissue in plants? a) Support and conduction b) Photosynthesis c) Water absorption d) Storage of water and nutrients

71. Which type of tissue is responsible for the production of essential oils in plants? a) Meristematic tissue b) Epidermal tissue c) Vascular tissue d) Secretory tissue

72. What is the function of the trichomes in plants? a) Support and conduction b) Photosynthesis c) Water absorption d) Protection and prevention of water loss

73. Which type of tissue is responsible for the formation of the apical bud in plants? a) Meristematic tissue b) Epidermal tissue c) Vascular tissue d) Ground tissue

74. What is the main role of the suberin in the cork cells of plants? a) Support and conduction b) Photosynthesis c) Protection and prevention of water loss d) Storage of nutrients

75. Which type of tissue is responsible for the production of resin in plants? a) Meristematic tissue b) Epidermal tissue c) Vascular tissue d) Secretory tissue

76. What is the function of the perivascular fibers in plants? a) Support and conduction b) Photosynthesis c) Storage of water and nutrients d) Mechanical strength and rigidity

77. Which type of tissue is responsible for the production of gums and mucilage in plants? a) Meristematic tissue b) Epidermal tissue c) Vascular tissue d) Secretory tissue

78. What is the main role of the endocarp in plants? a) Support and conduction b) Photosynthesis c) Protection of the seed d) Water absorption

79. Which type of tissue is responsible for the production of tannins in plants? a) Meristematic tissue b) Epidermal tissue c) Vascular tissue d) Secretory tissue

80. What is the function of the fibers in plants? a) Support and conduction b) Photosynthesis c) Water absorption d) Mechanical strength and rigidity

81. Which type of tissue is responsible for the production of latex in plants? a) Meristematic tissue b) Epidermal tissue c) Vascular tissue d) Laticiferous tissue

82. What is the main role of the laticiferous tissue in plants? a) Support and conduction b) Photosynthesis c) Storage of water and nutrients d) Latex production

83. Which type of tissue is responsible for the production of fibers in plants? a) Meristematic tissue b) Epidermal tissue c) Vascular tissue d) Sclerenchyma tissue

84. What is the function of the sclerenchyma tissue in plants? a) Support and conduction b) Photosynthesis c) Storage of water and nutrients d) Mechanical strength and rigidity

85. Which type of tissue is responsible for the production of spines and thorns in plants? a) Meristematic tissue b) Epidermal tissue c) Vascular tissue d) Sclerenchyma tissue

86. What is the main role of the companion cells in plants? a) Support and conduction b) Photosynthesis c) Storage of water and nutrients d) Assistance in phloem transport

87. Which type of tissue is responsible for the production of cork in plants? a) Meristematic tissue b) Epidermal tissue c) Vascular tissue d) Cork cambium

88. What is the function of the epidermal hairs (trichomes) in plants? a) Support and conduction b) Photosynthesis c) Water absorption d) Protection and prevention of water loss

89. Which type of tissue is responsible for the formation of the leaf veins in plants? a) Meristematic tissue b) Epidermal tissue c) Vascular tissue d) Ground tissue

90. What is the main role of the pericycle in plants? a) Support and conduction b) Photosynthesis c) Protection of the root d) Lateral root formation

91. Which type of tissue is responsible for the production of resin ducts in plants? a) Meristematic tissue b) Epidermal tissue c) Vascular tissue d) Secretory tissue

92. What is the function of the periderm in plants? a) Support and conduction b) Photosynthesis c) Protection and prevention of water loss d) Storage of nutrients

93. Which type of tissue is responsible for the production of nectar in plants? a) Meristematic tissue b) Epidermal tissue c) Vascular tissue d) Secretory tissue

94. What is the main role of the transfer cells in plants? a) Support and conduction b) Photosynthesis c) Water absorption d) Increased nutrient transport

95. Which type of tissue is responsible for the production of essential oils in plants? a) Meristematic tissue b) Epidermal tissue c) Vascular tissue d) Secretory tissue

96. What is the function of the glandular hairs (trichomes) in plants? a) Support and conduction b) Photosynthesis c) Water absorption d) Secretion of oils and other substances

97. Which type of tissue is responsible for the production of resin canals in plants? a) Meristematic tissue b) Epidermal tissue c) Vascular tissue d) Secretory tissue

98. What is the main role of the pith rays in plants? a) Support and conduction b) Photosynthesis c) Water absorption d) Storage of water and nutrients

99. Which type of tissue is responsible for the production of digestive enzymes in insectivorous plants? a) Meristematic tissue b) Epidermal tissue c) Vascular tissue d) Secretory tissue

100. What is the function of the nectaries in plants? a) Support and conduction b) Photosynthesis c) Water absorption d) Production of nectar

ANSWERS

1. a) Support and protection
2. c) Growth and cell division
3. c) Vascular tissue
4. c) Protection and prevention of water loss
5. b) Epidermal tissue
6. b) Support and conduction of water and nutrients
7. c) Vascular tissue
8. a) Support and flexibility
9. a) Meristematic tissue
10. a) Production of cork cells for protection
11. a) Meristematic tissue
12. d) Storage of food and nutrients
13. d) Ground tissue
14. c) Food storage and transport
15. c) Vascular tissue
16. c) Water absorption
17. d) Cork cambium
18. c) Food transport
19. c) Vascular cambium
20. d) Conduction of nutrients
21. d) Endodermis
22. a) Support and conduction
23. a) Meristematic tissue
24. c) Protection and prevention of water loss
25. c) Vascular cambium
26. c) Food transport
27. c) Parenchyma tissue
28. c) Growth in length

29. a) Meristematic tissue

30. c) Food storage and transport

31. d) Ground tissue

32. c) Growth in girth

33. d) Cork cambium

34. b) Epidermal tissue

35. c) Vascular tissue

36. c) Food transport

37. c) Vascular cambium

38. d) Conduction of nutrients

39. d) Endodermis

40. a) Support and conduction

41. c) Vascular cambium

42. c) Protection and prevention of water loss

43. c) Vascular tissue

44. c) Food transport

45. c) Parenchyma tissue

46. c) Growth in length

47. a) Meristematic tissue

48. b) Photosynthesis

49. d) Ground tissue

50. c) Growth in girth

51. d) Cork cambium

52. c) Food storage and nourishment

53. b) Epidermal tissue

54. c) Secondary growth

55. a) Meristematic tissue

56. a) Secretion of nectar and other substances

57. a) Meristematic tissue

58. d) Storage of water and nutrients

59. a) Meristematic tissue

60. d) Water absorption

61. d) Laticiferous tissue

62. d) Latex production

63. d) Sclerenchyma tissue

64. d) Mechanical strength and rigidity

65. d) Sclerenchyma tissue

66. d) Assistance in phloem transport

67. d) Cork cambium

68. c) Protection and prevention of water loss

69. b) Epidermal tissue

70. b) Photosynthesis

71. d) Secretory tissue

72. d) Protection and prevention of water loss

73. a) Meristematic tissue

74. c) Protection and prevention of water loss

75. d) Secretory tissue

76. d) Mechanical strength and rigidity

77. d) Secretory tissue

78. c) Protection of the seed

79. d) Secretory tissue

80. d) Mechanical strength and rigidity

81. d) Laticiferous tissue

82. d) Latex production

83. d) Sclerenchyma tissue

84. d) Mechanical strength and rigidity

85. d) Sclerenchyma tissue

86. d) Assistance in phloem transport

87. d) Cork cambium

88. d) Protection and prevention of water loss

89. c) Vascular tissue

90. d) Lateral root formation

91. d) Secretory tissue

92. c) Protection and prevention of water loss

93. d) Secretory tissue

94. d) Increased nutrient transport

95. d) Secretory tissue

96. d) Secretion of oils and other substances

97. d) Secretory tissue

98. d) Storage of water and nutrients

99. d) Secretory tissue

100. d) Production of nectar

ROOT ANATOMY & MORPHOLOGY

1. Which of the following is the primary function of roots in plants? a) Photosynthesis b) Storage of nutrients c) Reproduction d) Anchoring and absorption

2. What is the outermost layer of the root called? a) Cortex b) Epidermis c) Endodermis d) Pericycle

3. What is the region of actively dividing cells at the tip of a root called? a) Root cap b) Apical meristem c) Zone of elongation d) Zone of maturation

4. Which of the following is responsible for the absorption of water and minerals in roots? a) Root hairs b) Lateral roots c) Vascular bundles d) Periderm

5. What is the function of the root cap in plants? a) Protection of the root tip b) Water absorption c) Photosynthesis d) Storage of nutrients

6. Which type of root system consists of a single main root with smaller lateral roots? a) Fibrous root system b) Taproot system c) Adventitious root system d) Propagative root system

7. What is the term for the swelling at the base of a stem where the root system originates? a) Node b) Internode c) Bud d) Crown

8. Which of the following is NOT a function of root hairs? a) Anchoring the root in the soil b) Absorption of water and minerals c) Protection of the root tip d) Increasing the surface area for nutrient uptake

9. What is the innermost layer of the root called? a) Cortex b) Epidermis c) Endodermis d) Pericycle

10. Which type of root system is typically found in monocot plants? a) Fibrous root system b) Taproot system c) Adventitious root system d) Propagative root system

11. What is the term for the region of the root where lateral roots originate? a) Root cap b) Apical meristem c) Zone of elongation d) Zone of maturation

12. Which of the following is a characteristic of adventitious roots? a) They develop from the radicle of the embryo b) They arise from stems or leaves instead of the main root c) They have a taproot system d) They are found in fibrous root systems

13. What is the primary function of the endodermis in roots? a) Water absorption b) Nutrient storage c) Protection of the root tip d) Regulation of water and nutrient movement

14. Which type of root system is typically found in dicot plants? a) Fibrous root system b) Taproot system c) Adventitious root system d) Propagative root system

15. What is the term for the zone of the root where cells become specialized and differentiate? a) Root cap b) Apical meristem c) Zone of elongation d) Zone of maturation

16. Which of the following is responsible for secondary growth in roots? a) Vascular cambium b) Cork cambium c) Periderm d) Endodermis

17. What is the function of the pericycle in roots? a) Water absorption b) Nutrient storage c) Protection of the root tip d) Lateral root formation

18. Which of the following is NOT a characteristic of fibrous root systems? a) They have numerous fine roots of similar size b) They provide good anchorage in loose soil c) They are typically found in monocot plants d) They have a main taproot with smaller lateral roots

19. What is the term for the swollen region of the root that stores food and water? a) Root cap b) Apical meristem c) Zone of elongation d) Root bulb

20. Which of the following is a function of prop roots in plants? a) Nutrient absorption b) Photosynthesis c) Anchoring and support d) Water storage

21. Which type of root modification allows plants to breathe in waterlogged environments? a) Pneumatophores b) Contractile roots c) Buttress roots d) Haustorial roots

22. What is the term for the region of the root where cell elongation occurs? a) Root cap b) Apical meristem c) Zone of elongation d) Zone of maturation

23. Which of the following is responsible for the transport of water and nutrients in roots? a) Xylem b) Phloem c) Cambium d) Cork

24. What is the function of contractile roots in plants? a) Nutrient absorption b) Photosynthesis c) Anchoring and support d) Reduction of the root length

25. Which of the following is a characteristic of taproot systems? a) They have numerous fine roots of similar size b) They provide good anchorage in loose soil c) They are typically found in monocot plants d) They have a main root with smaller lateral roots

26. What is the term for the zone of the root where cells undergo rapid division and growth? a) Root cap b) Apical meristem c) Zone of elongation d) Zone of maturation

27. Which of the following root modifications allows plants to obtain nutrients from other plants? a) Pneumatophores b) Aerial roots c) Haustorial roots d) Prop roots

28. What is the function of root nodules in plants? a) Nutrient absorption b) Photosynthesis c) Nitrogen fixation d) Anchoring and support

29. Which of the following is responsible for the synthesis and secretion of root exudates? a) Epidermis b) Cortex c) Pericycle d) Rhizodermis

30. What is the term for the region of the root where cells become specialized in their functions? a) Root cap b) Apical meristem c) Zone of elongation d) Zone of maturation

31. Which of the following root modifications allows plants to climb and attach to supports? a) Pneumatophores b) Aerial roots c) Buttress roots d) Clinging roots

32. What is the function of root pressure in plants? a) Nutrient absorption b) Photosynthesis c) Water absorption d) Osmotic regulation

33. Which of the following is responsible for the storage of nutrients in roots? a) Epidermis b) Cortex c) Endodermis d) Pericycle

34. What is the term for the region of the root where cells undergo elongation and increase in size? a) Root cap b) Apical meristem c) Zone of elongation d) Zone of maturation

35. Which of the following is responsible for the radial transport of water and nutrients in roots? a) Xylem b) Phloem c) Cambium d) Cork

36. What is the function of aerial roots in plants? a) Nutrient absorption b) Photosynthesis c) Anchoring and support d) Gas exchange

37. Which of the following is responsible for the storage of water in roots? a) Epidermis b) Cortex c) Endodermis d) Pericycle

38. What is the term for the region of the root where cells differentiate into specific cell types? a) Root cap b) Apical meristem c) Zone of elongation d) Zone of maturation

39. Which of the following root modifications allows plants to obtain oxygen in waterlogged environments? a) Pneumatophores b) Contractile roots c) Buttress roots d) Prop roots

40. What is the function of buttress roots in plants? a) Nutrient absorption b) Photosynthesis c) Anchoring and support d) Water storage

41. Which of the following is responsible for the synthesis of root hormones and growth regulators? a) Root cap b) Apical meristem c) Zone of elongation d) Zone of maturation

42. What is the function of haustorial roots in plants? a) Nutrient absorption b) Photosynthesis c) Anchoring and support d) Parasitic attachment and nutrient absorption

43. Which of the following is responsible for the transport of sugars and organic molecules in roots? a) Xylem b) Phloem c) Cambium d) Cork

44. What is the term for the region of the root where cells mature and become fully functional? a) Root cap b) Apical meristem c) Zone of elongation d) Zone of maturation

45. Which of the following root modifications allows plants to absorb oxygen from the air? a) Pneumatophores b) Aerial roots c) Haustorial roots d) Prop roots

46. What is the function of clinging roots in plants? a) Nutrient absorption b) Photosynthesis c) Anchoring and support d) Clinging to surfaces for support

47. Which of the following is responsible for the uptake of water and minerals in roots? a) Epidermis b) Cortex c) Endodermis d) Pericycle

48. What is the term for the region of the root where cells divide and give rise to new cells? a) Root cap b) Apical meristem c) Zone of elongation d) Zone of maturation

49. Which of the following is responsible for the transport of water and dissolved minerals from the roots to the rest of the plant? a) Xylem b) Phloem c) Cambium d) Cork

50. What is the function of pneumatophores in plants? a) Nutrient absorption b) Photosynthesis c) Anchoring and support d) Gas exchange in waterlogged environments

51. Which of the following root modifications allows plants to provide additional support in shallow soils? a) Pneumatophores b) Aerial roots c) Buttress roots d) Contractile roots

52. What is the function of rhizomes in plants? a) Nutrient absorption b) Photosynthesis c) Anchoring and support d) Storage and vegetative propagation

53. Which of the following is responsible for the transport of carbohydrates and hormones in roots? a) Xylem b) Phloem c) Cambium d) Cork

54. What is the term for the region of the root where cells differentiate into specialized tissues? a) Root cap b) Apical meristem c) Zone of elongation d) Zone of maturation

55. Which of the following root modifications allows plants to obtain nutrients from decaying organic matter? a) Pneumatophores b) Aerial roots c) Haustorial roots d) Storage roots

56. What is the function of storage roots in plants? a) Nutrient absorption b) Photosynthesis c) Anchoring and support d) Storage of carbohydrates and nutrients

57. Which of the following is responsible for the lateral growth and increase in girth of roots? a) Vascular cambium b) Cork cambium c) Periderm d) Endodermis

58. What is the term for the region of the root where cells mature and acquire their specific functions? a) Root cap b) Apical meristem c) Zone of elongation d) Zone of maturation

59. Which of the following root modifications allows plants to obtain nutrients from other plants? a) Pneumatophores b) Aerial roots c) Haustorial roots d) Prop roots

60. What is the function of taproots in plants? a) Nutrient absorption b) Photosynthesis c) Anchoring and support d) Storage of water and nutrients

61. Which of the following is responsible for the transport of water and nutrients between the root and the shoot system? a) Xylem b) Phloem c) Cambium d) Cork

62. What is the term for the region of the root where cells elongate and increase in length? a) Root cap b) Apical meristem c) Zone of elongation d) Zone of maturation

63. Which of the following root modifications allows plants to store water and nutrients in arid environments? a) Pneumatophores b) Aerial roots c) Contractile roots d) Storage roots

64. What is the function of stilt roots in plants? a) Nutrient absorption b) Photosynthesis c) Anchoring and support d) Propagation through runners

65. Which of the following is responsible for the radial transport of carbohydrates and hormones in roots? a) Xylem b) Phloem c) Cambium d) Cork

66. What is the term for the region of the root where cells differentiate and acquire specific functions? a) Root cap b) Apical meristem c) Zone of elongation d) Zone of maturation

67. Which of the following root modifications allows plants to anchor themselves in shallow or unstable soils? a) Pneumatophores b) Aerial roots c) Buttress roots d) Prop roots

68. What is the function of contractile roots in plants? a) Nutrient absorption b) Photosynthesis c) Anchoring and support d) Reduction of the root length

69. Which of the following is responsible for the storage of carbohydrates and nutrients in roots? a) Epidermis b) Cortex c) Endodermis d) Pericycle

70. What is the term for the region of the root where cells undergo elongation and increase in size? a) Root cap b) Apical meristem c) Zone of elongation d) Zone of maturation

71. Which of the following root modifications allows plants to obtain oxygen from the air? a) Pneumatophores b) Aerial roots c) Haustorial roots d) Prop roots

72. What is the function of prop roots in plants? a) Nutrient absorption b) Photosynthesis c) Anchoring and support d) Water storage

73. Which of the following is responsible for the transport of water and dissolved minerals from the roots to the rest of the plant? a) Xylem b) Phloem c) Cambium d) Cork

74. What is the term for the region of the root where cells mature and become fully functional? a) Root cap b) Apical meristem c) Zone of elongation d) Zone of maturation

75. Which of the following root modifications allows plants to obtain nutrients from decaying organic matter? a) Pneumatophores b) Aerial roots c) Haustorial roots d) Storage roots

76. What is the function of buttress roots in plants? a) Nutrient absorption b) Photosynthesis c) Anchoring and support d) Water storage

77. Which of the following is responsible for the transport of carbohydrates and hormones in roots? a) Xylem b) Phloem c) Cambium d) Cork

78. What is the term for the region of the root where cells differentiate into specialized tissues? a) Root cap b) Apical meristem c) Zone of elongation d) Zone of maturation

79. Which of the following root modifications allows plants to obtain nutrients from other plants? a) Pneumatophores b) Aerial roots c) Haustorial roots d) Prop roots

80. What is the function of storage roots in plants? a) Nutrient absorption b) Photosynthesis c) Anchoring and support d) Storage of carbohydrates and nutrients

81. Which of the following is responsible for the lateral growth and increase in girth of roots? a) Vascular cambium b) Cork cambium c) Periderm d) Endodermis

82. What is the term for the region of the root where cells mature and acquire their specific functions? a) Root cap b) Apical meristem c) Zone of elongation d) Zone of maturation

83. Which of the following root modifications allows plants to obtain nutrients from other plants? a) Pneumatophores b) Aerial roots c) Haustorial roots d) Prop roots

84. What is the function of taproots in plants? a) Nutrient absorption b) Photosynthesis c) Anchoring and support d) Storage of water and nutrients

85. Which of the following is responsible for the transport of water and nutrients between the root and the shoot system? a) Xylem b) Phloem c) Cambium d) Cork

86. What is the term for the region of the root where cells elongate and increase in length? a) Root cap b) Apical meristem c) Zone of elongation d) Zone of maturation

87. Which of the following root modifications allows plants to store water and nutrients in arid environments? a) Pneumatophores b) Aerial roots c) Contractile roots d) Storage roots

88. What is the function of stilt roots in plants? a) Nutrient absorption b) Photosynthesis c) Anchoring and support d) Propagation through runners

89. Which of the following is responsible for the radial transport of carbohydrates and hormones in roots? a) Xylem b) Phloem c) Cambium d) Cork

90. What is the term for the region of the root where cells differentiate and acquire specific functions? a) Root cap b) Apical meristem c) Zone of elongation d) Zone of maturation

91. Which of the following root modifications allows plants to anchor themselves in shallow or unstable soils? a) Pneumatophores b) Aerial roots c) Buttress roots d) Prop roots

92. What is the function of contractile roots in plants? a) Nutrient absorption b) Photosynthesis c) Anchoring and support d) Reduction of the root length

93. Which of the following is responsible for the storage of carbohydrates and nutrients in roots? a) Epidermis b) Cortex c) Endodermis d) Pericycle

94. What is the term for the region of the root where cells undergo elongation and increase in size? a) Root cap b) Apical meristem c) Zone of elongation d) Zone of maturation

95. Which of the following root modifications allows plants to obtain oxygen from the air? a) Pneumatophores b) Aerial roots c) Haustorial roots d) Prop roots

96. What is the function of prop roots in plants? a) Nutrient absorption b) Photosynthesis c) Anchoring and support d) Water storage

97. Which of the following is responsible for the transport of water and dissolved minerals from the roots to the rest of the plant? a) Xylem b) Phloem c) Cambium d) Cork

98. What is the term for the region of the root where cells mature and become fully functional? a) Root cap b) Apical meristem c) Zone of elongation d) Zone of maturation

99. Which of the following root modifications allows plants to obtain nutrients from decaying organic matter? a) Pneumatophores b) Aerial roots c) Haustorial roots d) Storage roots

100. What is the function of buttress roots in plants? a) Nutrient absorption b) Photosynthesis c) Anchoring and support d) Water storage

ANSWERS

1. d) Anchoring and absorption
2. b) Epidermis
3. b) Apical meristem
4. a) Root hairs
5. a) Protection of the root tip
6. b) Taproot system
7. d) Crown
8. c) Protection of the root tip
9. c) Endodermis
10. a) Fibrous root system
11. b) Apical meristem
12. b) They arise from stems or leaves instead of the main root
13. d) Regulation of water and nutrient movement
14. b) Taproot system
15. d) Zone of maturation
16. a) Vascular cambium
17. d) Lateral root formation
18. d) They have a main taproot with smaller lateral roots
19. d) Root bulb
20. c) Anchoring and support
21. a) Pneumatophores
22. c) Zone of elongation
23. a) Xylem
24. d) Reduction of the root length
25. b) They have a main root with smaller lateral roots
26. b) Apical meristem
27. d) Prop roots
28. c) Nitrogen fixation

29. b) Cortex

30. d) Zone of maturation

31. b) Aerial roots

32. c) Water absorption

33. b) Phloem

34. b) Apical meristem

35. a) Xylem

36. b) Photosynthesis

37. b) Cortex

38. c) Zone of elongation

39. a) Pneumatophores

40. d) Storage of carbohydrates and nutrients

41. b) Apical meristem

42. c) Anchoring and support

43. b) Phloem

44. d) Zone of maturation

45. a) Pneumatophores

46. d) Clinging to surfaces for support

47. a) Epidermis

48. b) Apical meristem

49. a) Xylem

50. b) Photosynthesis

51. c) Buttress roots

52. d) Storage and vegetative propagation

53. b) Phloem

54. d) Zone of maturation

55. c) Haustorial roots

56. d) Storage of water and nutrients

57. a) Vascular cambium

58. d) Zone of maturation

59. c) Haustorial roots

60. c) Anchoring and support

61. b) Phloem

62. c) Zone of elongation

63. a) Pneumatophores

64. c) Anchoring and support

65. b) Phloem

66. d) Zone of maturation

67. c) Buttress roots

68. d) Reduction of the root length

69. b) Cortex

70. c) Zone of elongation

71. b) Aerial roots

72. c) Anchoring and support

73. a) Xylem

74. d) Zone of maturation

75. d) Storage roots

76. c) Anchoring and support

77. b) Phloem

78. d) Zone of maturation

79. c) Haustorial roots

80. d) Storage of carbohydrates and nutrients

81. a) Vascular cambium

82. d) Zone of maturation

83. c) Haustorial roots

84. b) Anchoring and support

85. a) Xylem

86. c) Zone of elongation

87. d) Storage roots

88. c) Anchoring and support

89. b) Phloem

90. d) Zone of maturation

91. c) Buttress roots

92. c) Anchoring and support

93. b) Cortex

94. c) Zone of elongation

95. b) Aerial roots

96. c) Anchoring and support

97. a) Xylem

98. d) Zone of maturation

99. d) Storage roots

100.c) Anchoring and support

STEM ANATOMY & MORPHOLOGY

1. Which of the following is NOT a function of stems in plants? a) Support b) Photosynthesis c) Transport of water and nutrients d) Storage

2. What is the region of a stem where leaves are attached called? a) Node b) Internode c) Bud d) Apex

3. Which type of stem modification allows plants to climb and attach to supports? a) Stolons b) Rhizomes c) Thorns d) Tendrils

4. What is the term for the protective outer layer of a stem? a) Cortex b) Epidermis c) Endodermis d) Pericycle

5. What is the function of lenticels in stems? a) Gas exchange b) Water absorption c) Photosynthesis d) Nutrient storage

6. Which of the following is responsible for the primary growth of stems? a) Vascular cambium b) Apical meristem c) Cork cambium d) Lateral meristem

7. What is the term for the tissue that transports water and nutrients in stems? a) Xylem b) Phloem c) Cambium d) Cork

8. Which type of stem modification allows plants to store water and nutrients? a) Rhizomes b) Tubers c) Bulbs d) Corms

9. Which of the following is NOT a characteristic of herbaceous stems? a) They are soft and flexible b) They undergo secondary growth c) They contain chlorophyll for photosynthesis d) They have a shorter lifespan compared to woody stems

10. What is the function of thorns in stems? a) Water absorption b) Protection against herbivores c) Anchoring and support d) Storage of nutrients

11. Which of the following is responsible for the radial growth of stems? a) Vascular cambium b) Apical meristem c) Cork cambium d) Lateral meristem

12. What is the term for the swollen base of a stem that provides support and stores nutrients? a) Node b) Internode c) Bud d) Bulb

13. Which type of stem modification allows plants to spread and form new individuals? a) Stolons b) Rhizomes c) Runners d) Tendrils

14. What is the function of pith in stems? a) Water absorption b) Nutrient storage c) Gas exchange d) Anchoring and support

15. Which of the following is responsible for the transport of sugars and organic molecules in stems? a) Xylem b) Phloem c) Cambium d) Cork

16. What is the term for the swollen underground stem that stores food and nutrients? a) Rhizome b) Tuber c) Bulb d) Corm

17. Which type of stem modification allows plants to store water and carbohydrates? a) Stolons b) Rhizomes c) Tubers d) Corms

18. What is the function of tendrils in stems? a) Water absorption b) Protection against herbivores c) Anchoring and support d) Climbing and attachment

19. Which of the following is responsible for the transport of water and minerals from the roots to the stems? a) Xylem b) Phloem c) Cambium d) Cork

20. What is the term for the swollen underground stem that consists of fleshy storage leaves? a) Rhizome b) Tuber c) Bulb d) Corm

21. Which type of stem modification allows plants to produce new shoots and roots at the nodes? a) Stolons b) Rhizomes c) Tubers d) Runners

22. What is the function of bulbs in stems? a) Water absorption b) Nutrient storage c) Gas exchange d) Climbing and attachment

23. Which of the following is responsible for the synthesis and secretion of stem exudates? a) Epidermis b) Cortex c) Pericycle d) Rhizome

24. What is the term for the region of the stem where cells divide and give rise to new cells? a) Vascular cambium b) Apical meristem c) Cork cambium d) Lateral meristem

25. Which type of stem modification allows plants to produce new plants from the base of the stem? a) Stolons b) Rhizomes c) Bulbs d) Corms

26. What is the function of stolons in stems? a) Water absorption b) Protection against herbivores c) Anchoring and support d) Asexual reproduction

27. Which of the following is responsible for the synthesis of stem hormones and growth regulators? a) Epidermis b) Cortex c) Pericycle d) Rhizome

28. What is the term for the region of the stem where cells elongate and increase in size?
   a) Vascular cambium b) Apical meristem c) Cork cambium d) Lateral meristem

29. Which type of stem modification allows plants to produce new shoots from underground stems? a) Stolons b) Rhizomes c) Tubers d) Corms

30. What is the function of cork cambium in stems? a) Water absorption b) Protection against herbivores c) Anchoring and support d) Production of cork cells forthe outer protective layer

31. Which of the following is responsible for the transport of water and minerals in stems?
   a) Xylem b) Phloem c) Cambium d) Cork

32. What is the term for the region of the stem where cells differentiate into specialized tissues? a) Vascular cambium b) Apical meristem c) Cork cambium d) Lateral meristem

33. Which type of stem modification allows plants to produce new shoots from underground buds? a) Stolons b) Rhizomes c) Tubers d) Corms

34. What is the function of cork in stems? a) Water absorption b) Nutrient storage c) Gas exchange d) Protection and insulation

35. Which of the following is responsible for the transport of sugars and organic molecules in stems? a) Xylem b) Phloem c) Cambium d) Cork

36. What is the term for the region of the stem where cells mature and become fully functional? a) Vascular cambium b) Apical meristem c) Cork cambium d) Lateral meristem

37. Which type of stem modification allows plants to produce new plants from the base of the stem? a) Stolons b) Rhizomes c) Bulbs d) Corms

38. What is the function of tubers in stems? a) Water absorption b) Nutrient storage c) Gas exchange d) Climbing and attachment

39. Which of the following is responsible for the synthesis and secretion of stem exudates?
   a) Epidermis b) Cortex c) Pericycle d) Rhizome

40. What is the term for the region of the stem where cells divide and give rise to new cells? a) Vascular cambium b) Apical meristem c) Cork cambium d) Lateral meristem

41. Which type of stem modification allows plants to produce new plants from underground stems? a) Stolons b) Rhizomes c) Bulbs d) Corms

42. What is the function of stolons in stems? a) Water absorption b) Protection against herbivores c) Anchoring and support d) Asexual reproduction

43. Which of the following is responsible for the synthesis of stem hormones and growth regulators? a) Epidermis b) Cortex c) Pericycle d) Rhizome

44. What is the term for the region of the stem where cells elongate and increase in size? a) Vascular cambium b) Apical meristem c) Cork cambium d) Lateral meristem

45. Which type of stem modification allows plants to produce new shoots from underground stems? a) Stolons b) Rhizomes c) Bulbs d) Corms

46. What is the function of cork cambium in stems? a) Water absorption b) Protection against herbivores c) Anchoring and support d) Production of cork cells for the outer protective layer

47. Which of the following is responsible for the transport of water and minerals in stems? a) Xylem b) Phloem c) Cambium d) Cork

48. What is the term for the region of the stem where cells differentiate into specialized tissues? a) Vascular cambium b) Apical meristem c) Cork cambium d) Lateral meristem

49. Which type of stem modification allows plants to produce new shoots from underground buds? a) Stolons b) Rhizomes c) Bulbs d) Corms

50. What is the function of cork in stems? a) Water absorption b) Nutrient storage c) Gas exchange d) Protection and insulation

51. Which of the following is responsible for the transport of sugars and organic molecules in stems? a) Xylem b) Phloem c) Cambium d) Cork

52. What is the term for the region of the stem where cells mature and become fully functional? a) Vascular cambium b) Apical meristem c) Cork cambium d) Lateral meristem

53. Which type of stem modification allows plants to produce new plants from the base of the stem? a) Stolons b) Rhizomes c) Bulbs d) Corms

54. What is the function of tubers in stems? a) Water absorption b) Nutrient storage c) Gas exchange d) Climbing and attachment

55. Which of the following is responsible for the synthesis and secretion of stem exudates? a) Epidermis b) Cortex c) Pericycle d) Rhizome

56. What is the term for the region of the stem where cells divide and give rise to new cells? a) Vascular cambium b) Apical meristem c) Cork cambium d) Lateral meristem

57. Which type of stem modification allows plants to produce new plants from underground stems? a) Stolons b) Rhizomes c) Bulbs d) Corms

58. What is the function of stolons in stems? a) Water absorption b) Protection against herbivores c) Anchoring and support d) Asexual reproduction

59. Which of the following is responsible for the synthesis of stem hormones and growth regulators? a) Epidermis b) Cortex c) Pericycle d) Rhizome

60. What is the term for the region of the stem where cells elongate and increase in size? a) Vascular cambium b) Apical meristem c) Cork cambium d) Lateral meristem

61. Which type of stem modification allows plants to produce new shoots from underground stems? a) Stolons b) Rhizomes c) Bulbs d) Corms

62. What is the function of cork cambium in stems? a) Water absorption b) Protection against herbivores c) Anchoring and support d) Production of cork cells for the outer protective layer

63. Which of the following is responsible for the transport of water and minerals in stems? a) Xylem b) Phloem c) Cambium d) Cork

64. What is the term for the region of the stem where cells differentiate into specialized tissues? a) Vascular cambium b) Apical meristem c) Cork cambium d) Lateral meristem

65. Which type of stem modification allows plants to produce new shoots from underground buds? a) Stolons b) Rhizomes c) Bulbs d) Corms

66. What is the function of cork in stems? a) Water absorption b) Nutrient storage c) Gas exchange d) Protection and insulation

67. Which of the following is responsible for the transport of sugars and organic molecules in stems? a) Xylem b) Phloem c) Cambium d) Cork

68. What is the term for the region of the stem where cells mature and become fully functional? a) Vascular cambium b) Apical meristem c) Cork cambium d) Lateral meristem

69. Which type of stem modification allows plants to produce new plants from the base of the stem? a) Stolons b) Rhizomes c) Bulbs d) Corms

70. What is the function of cork in stems? a) Water absorption b) Nutrient storage c) Gas exchange d) Protection and insulation

71. Which of the following is responsible for the transport of sugars and organic molecules in stems? a) Xylem b) Phloem c) Cambium d) Cork

72. What is the term for the region of the stem where cells mature and become fully functional? a) Vascular cambium b) Apical meristem c) Cork cambium d) Lateral meristem

73. Which type of stem modification allows plants to produce new plants from underground stems? a) Stolons b) Rhizomes c) Bulbs d) Corms

74. What is the function of tubers in stems? a) Water absorption b) Nutrient storage c) Gas exchange d) Climbing and attachment

75. Which of the following is responsible for the synthesis and secretion of stem exudates? a) Epidermis b) Cortex c) Pericycle d) Rhizome

76. What is the term for the region of the stem where cells divide and give rise to new cells? a) Vascular cambium b) Apical meristem c) Cork cambium d) Lateral meristem

77. Which type of stem modification allows plants to produce new shoots from underground stems? a) Stolons b) Rhizomes c) Bulbs d) Corms

78. What is the function of stolons in stems? a) Water absorption b) Protection against herbivores c) Anchoring and support d) Asexual reproduction

79. Which of the following is responsible for the synthesis of stem hormones and growth regulators? a) Epidermis b) Cortex c) Pericycle d) Rhizome

80. What is the term for the region of the stem where cells elongate and increase in size? a) Vascular cambium b) Apical meristem c) Cork cambium d) Lateral meristem

81. Which type of stem modification allows plants to produce new shoots from underground stems? a) Stolons b) Rhizomes c) Bulbs d) Corms

82. What is the function of cork cambium in stems? a) Water absorption b) Protection against herbivores c) Anchoring and support d) Production of cork cells for the outer protective layer

83. Which of the following is responsible for the transport of water and minerals in stems? a) Xylem b) Phloem c) Cambium d) Cork

84. What is the term for the region of the stem where cells differentiate into specialized tissues? a) Vascular cambium b) Apical meristem c) Cork cambium d) Lateral meristem

85. Which type of stem modification allows plants to produce new shoots from underground buds? a) Stolons b) Rhizomes c) Bulbs d) Corms

86. What is the function of cork in stems? a) Water absorption b) Nutrient storage c) Gas exchange d) Protection and insulation

87. Which of the following is responsible for the transport of sugars and organic molecules in stems? a) Xylem b) Phloem c) Cambiumd) Cork

88. What is the term for the region of the stem where cells mature and become fully functional? a) Vascular cambium b) Apical meristem c) Cork cambium d) Lateral meristem

89. Which type of stem modification allows plants to produce new plants from the base of the stem? a) Stolons b) Rhizomes c) Bulbs d) Corms

90. What is the function of cork in stems? a) Water absorption b) Nutrient storage c) Gas exchange d) Protection and insulation

91. Which of the following is responsible for the transport of sugars and organic molecules in stems? a) Xylem b) Phloem c) Cambium d) Cork

92. What is the term for the region of the stem where cells mature and become fully functional? a) Vascular cambium b) Apical meristem c) Cork cambium d) Lateral meristem

93. Which type of stem modification allows plants to produce new plants from underground stems? a) Stolons b) Rhizomes c) Bulbs d) Corms

94. What is the function of tubers in stems? a) Water absorption b) Nutrient storage c) Gas exchange d) Climbing and attachment

95. Which of the following is responsible for the synthesis and secretion of stem exudates? a) Epidermis b) Cortex c) Pericycle d) Rhizome

96. What is the term for the region of the stem where cells divide and give rise to new cells? a) Vascular cambium b) Apical meristem c) Cork cambium d) Lateral meristem

97. Which type of stem modification allows plants to produce new shoots from underground stems? a) Stolons b) Rhizomes c) Bulbs d) Corms

98. What is the function of stolons in stems? a) Water absorption b) Protection against herbivores c) Anchoring and support d) Asexual reproduction

99. Which of the following is responsible for the synthesis of stem hormones and growth regulators? a) Epidermis b) Cortex c) Pericycle d) Rhizome

100.What is the term for the region of the stem where cells elongate and increase in size? a) Vascular cambium b) Apical meristem c) Cork cambium d) Lateral meristem

ANSWERS

1. b) Photosynthesis
2. a) Node
3. d) Tendrils
4. b) Epidermis
5. a) Gas exchange
6. b) Apical meristem
7. a) Xylem
8. c) Bulbs
9. b) They undergo secondary growth
10. b) Protection against herbivores
11. a) Vascular cambium
12. d) Bulb
13. a) Stolons
14. b) Nutrient storage
15. b) Phloem
16. b) Tuber
17. c) Tubers
18. c) Anchoring and support
19. a) Xylem
20. c) Bulb
21. d) Runners
22. b) Nutrient storage
23. a) Epidermis
24. b) Apical meristem
25. b) Rhizomes
26. d) Asexual reproduction
27. d) Rhizome
28. c) Cork cambium

29. b) Rhizomes

30. d) Production of cork cells for the outer protective layer

31. a) Xylem

32. b) Apical meristem

33. b) Rhizomes

34. a) Water absorption

35. b) Phloem

36. d) Lateral meristem

37. c) Bulbs

38. b) Protection against herbivores

39. d) Rhizome

40. b) Apical meristem

41. d) Corms

42. c) Anchoring and support

43. b) Cortex

44. a) Vascular cambium

45. b) Rhizomes

46. d) Protection and insulation

47. a) Xylem

48. b) Apical meristem

49. a) Xylem

50. d) Protection and insulation

51. b) Phloem

52. d) Lateral meristem

53. b) Rhizomes

54. b) Nutrient storage

55. d) Rhizome

56. b) Apical meristem

57. a) Xylem

58. d) Lateral meristem

59. b) Rhizomes

60. c) Anchoring and support

61. b) Phloem

62. c) Cork cambium

63. a) Xylem

64. b) Apical meristem

65. b) Rhizomes

66. c) Gas exchange

67. b) Phloem

68. d) Lateral meristem

69. b) Rhizomes

70. b) Nutrient storage

71. b) Phloem

72. d) Lateral meristem

73. b) Rhizomes

74. b) Nutrient storage

75. d) Rhizome

76. b) Apical meristem

77. b) Rhizomes

78. c) Anchoring and support

79. d) Rhizome

80. c) Cork cambium

81. b) Phloem

82. c) Cork cambium

83. a) Xylem

84. b) Apical meristem

85. b) Rhizomes

86. d) Protection and insulation

87. b) Phloem

88. d) Lateral meristem

89. b) Rhizomes

90. b) Nutrient storage

91. b) Phloem

92. d) Lateral meristem

93. b) Rhizomes

94. b) Nutrient storage

95. d) Rhizome

96. b) Apical meristem

97. b) Rhizomes

98. d) Asexual reproduction

99. d) Rhizome

100.b) Apical meristem

LEAF ANATOMY & MORPHOLOGY

1. The waxy, protective layer on the outer surface of a leaf is called the: a) Cuticle b) Stomata c) Mesophyll d) Epidermis

2. Which of the following is NOT a primary function of leaves in plants? a) Photosynthesis b) Transpiration c) Gas exchange d) Water absorption

3. The tissue responsible for the transport of water and nutrients in leaves is called: a) Xylem b) Phloem c) Epidermis d) Mesophyll

4. The tiny openings on the surface of leaves that regulate gas exchange are called: a) Stomata b) Cuticles c) Trichomes d) Guard cells

5. The main photosynthetic tissue in leaves is the: a) Epidermis b) Stomata c) Mesophyll d) Vascular bundle

6. Which of the following leaf types is characterized by having a single undivided blade? a) Simple leaf b) Compound leaf c) Palmate leaf d) Pinnate leaf

7. The vascular tissue responsible for the transport of sugars and organic molecules in leaves is called: a) Xylem b) Phloem c) Epidermis d) Mesophyll

8. The specialized cells surrounding the stomata that control their opening and closing are called: a) Cuticles b) Trichomes c) Guard cells d) Sclerenchyma cells

9. Which of the following leaf adaptations helps reduce water loss in hot and dry environments? a) Stomatal crypts b) Hairs on the leaf surface c) Sunken stomata d) Large surface area

10. The process by which leaves lose water vapor to the atmosphere is called: a) Transpiration b) Photosynthesis c) Respiration d) Evaporation

11. The arrangement of leaves along the stem is referred to as: a) Phyllotaxy b) Epidermis c) Mesophyll d) Vascular bundle

12. The spongy layer of mesophyll cells in leaves is involved in: a) Gas exchange b) Water absorption c) Nutrient storage d) Photosynthesis

13. The leaf modification that helps plants capture and digest insects is called: a) Tendril b) Spine c) Pitcher d) Bract

14. The leaf structure that connects the blade to the stem is called the: a) Petiole b) Midrib c) Stipule d) Margin

15. The region of the leaf where most photosynthesis occurs is the: a) Vein b) Epidermis c) Stomata d) Mesophyll

16. The leaf modification that helps reduce water loss in desert plants is called: a) Spine b) Scale c) Bract d) Pubescence

17. The primary pigment responsible for capturing light energy during photosynthesis is: a) Chlorophyll b) Carotenoid c) Anthocyanin d) Xanthophyll

18. The leaf adaptation that allows plants to survive in low-light environments is called: a) Tendril b) Bract c) Shade leaf d) Sun leaf

19. The leaf vein arrangement characterized by parallel veins is called: a) Palmate venation b) Pinnate venation c) Parallel venation d) Netted venation

20. The leaf modification that stores water in desert plants is called: a) Tendril b) Bract c) Succulent leaf d) Spine

21. The outermost layer of cells in a leaf is called the: a) Epidermis b) Mesophyll c) Cuticle d) Vascular bundle

22. The leaf adaptation that helps plants climb and attach to supports is called: a) Tendril b) Spine c) Bract d) Pubescence

23. The leaf vein arrangement characterized by a central midrib and smaller branching veins is called: a) Palmate venation b) Pinnate venation c) Parallel venation d) Netted venation

24. The leaf modification that protects plant buds when they are young is called a: a) Tendril b) Spine c) Bud scale d) Pubescence

25. The opening and closing of stomata is primarily regulated by changes in the concentration of: a) Oxygen b) Carbon dioxide c) Water vapor d) Potassium ions

26. The leaf adaptation that helps plants capture and digest small organisms is called: a) Tendril b) Spine c) Pitcher d) Bract

27. The leaf venation pattern characterized by several main veins branching off from a single point is called: a) Palmate venation b) Pinnate venation c) Parallel venation d) Netted venation

28. The leaf modification that helps plants attract pollinators is called a: a) Tendril b) Spine c) Bract d) Pubescence

29. The leaf adaptation that helps plants float on water is called: a) Tendril b) Spine c) Bract d) Air bladder

30. The process by which plants close their stomata to conserve water is called: a) Transpiration b) Photosynthesis c) Absorption d) Stomatal closure

31. The leaf modification that helps reduce water loss through its dense covering of fine hairs is called: a) Tendril b) Trichome c) Bract d) Pubescence

32. The leaf venation pattern characterized by a network of interconnected veins is called: a) Palmate venation b) Pinnate venation c) Parallel venation d) Netted venation

33. The specialized cells within the leaf epidermis that secrete a waxy substance are called: a) Trichomes b) Stomata c) Guard cells d) Cuticles

34. The leaf adaptation that helps plants trap and digest prey is called: a) Tendril b) Spine c) Pitcher d) Bract

35. The leaf structure that functions as a support for the leaf and transports nutrients is the: a) Petiole b) Midrib c) Stipule d) Margin

36. The leaf modification that stores water in specialized cells, often in desert plants, is called: a) Tendril b) Bulb c) Succulent leaf d) Spine

37. The leaf venation pattern characterized by several main veins running parallel to each other is called: a) Palmate venation b) Pinnate venation c) Parallel venation d) Netted venation

38. The specialized cells within the leaf epidermis that regulate the opening and closing of stomata are called: a) Trichomes b) Cuticles c) Guard cells d) Sclerenchyma cells

39. The leaf adaptation that helps plants reduce herbivory by having sharp, pointed structures is called a: a) Tendril b) Spine c) Bract d) Pubescence

40. The layer of cells located between the upper and lower epidermis in a leaf is called the: a) Palisade mesophyll b) Spongy mesophyll c) Xylem d) Phloem

41. The leaf modification that functions as a climbing or anchoring structure is called a: a) Tendril b) Spine c) Bract d) Pubescence

42. The leaf venation pattern characterized by several main veins branching out from a central point is called: a) Palmate venation b) Pinnate venation c) Parallel venation d) Netted venation

43. The specialized cells within the leaf mesophyll that contain chloroplasts and are responsible for photosynthesis are called: a) Trichomes b) Stomata c) Guard cells d) Chloroplasts

44. The leaf adaptation that helps plants reduce water loss by curling or folding the leaf is called: a) Tendril b) Spine c) Bract d) Leaf rolling

45. The leaf structure that connects the blade to the stem and allows for movement and orientation is the: a) Petiole b) Midrib c) Stipule d) Margin

46. The leaf modification that functions as a protective structure for young leaves or buds is called a: a) Tendril b) Spine c) Bract d) Pubescence

47. The layer of cells located just below the upper epidermis in a leaf, involved in photosynthesis, is the: a) Palisade mesophyll b) Spongy mesophyll c) Xylem d) Phloem

48. The leaf adaptation that helps plants repel herbivores with chemicals or toxins is called: a) Tendril b) Spine c) Bract d) Pubescence

49. The leaf venation pattern characterized by one main vein with smaller veins branching out from it is called: a) Palmate venation b) Pinnate venation c) Parallel venation d) Netted venation

50. The specialized cells within the leaf epidermis that help reduce water loss through their protective covering are called: a) Trichomes b) Cuticles c) Guard cells d) Sclerenchyma cells

51. The leaf adaptation that helps plants conserve water by reducing the surface area exposed to the environment is called: a) Tendril b) Spine c) Bract d) Leaf folding

52. The leaf structure that surrounds the base of a leaf where it attaches to the stem is called the: a) Petiole b) Midrib c) Stipule d) Margin

53. The leaf modification that functions as a protective covering for the growing tip of a young shoot is called a: a) Tendril b) Spine c) Bract d) Bud scale

54. The leaf venation pattern characterized by a single main vein running down the center of the leaf is called: a) Palmate venation b) Pinnate venation c) Parallel venation d) Netted venation

55. The specialized cells within the leaf epidermis that help reduce water loss by reflecting sunlight are called: a) Trichomes b) Cuticles c) Guard cells d) Sclerenchyma cells

56. The leaf adaptation that helps plants capture and retain water, particularly in arid environments, is called: a) Tendril b) Spine c) Bract d) Succulent leaf

57. The layer of cells located just below the lower epidermis in a leaf, involved in gas exchange and containing air spaces, is the: a) Palisade mesophyll b) Spongy mesophyll c) Xylem d) Phloem

58. The leaf modification that functions as a protective structure against herbivores and helps reduce water loss is called a: a) Tendril b) Spine c) Bract d) Pubescence

59. The leaf venation pattern characterized by several main veins radiating outward from a single point is called: a) Palmate venation b) Pinnate venation c) Parallel venation d) Netted venation

60. The specialized cells within the leaf mesophyll that are responsible for gas exchange and often contain chloroplasts are called: a) Trichomes b) Stomata c) Guard cells d) Parenchyma cells

ANSWERS

1. a) Cuticle
2. d) Water absorption
3. b) Phloem
4. a) Stomata
5. c) Mesophyll
6. a) Simple leaf
7. b) Phloem
8. c) Guard cells
9. c) Sunken stomata
10. a) Transpiration
11. a) Phyllotaxy
12. a) Gas exchange
13. c) Pitcher
14. a) Petiole
15. d) Mesophyll
16. b) Scale
17. a) Chlorophyll
18. c) Shade leaf
19. c) Parallel venation
20. c) Succulent leaf
21. a) Epidermis
22. a) Tendril
23. b) Pinnate venation
24. c) Bud scale
25. d) Potassium ions
26. c) Pitcher
27. a) Palmate venation
28. c) Bract
29. d) Air bladder
30. d) Stomatal closure

31. b) Trichome
32. d) Netted venation
33. d) Cuticles
34. c) Pitcher
35. b) Midrib
36. c) Succulent leaf
37. c) Parallel venation
38. c) Guard cells
39. b) Spine
40. b) Spongy mesophyll
41. a) Tendril
42. a) Palmate venation
43. d) Chloroplasts
44. d) Leaf rolling
45. a) Petiole
46. c) Bract
47. a) Palisade mesophyll
48. b) Spine
49. b) Pinnate venation
50. a) Trichomes
51. 51. d) Leaf folding
52. c) Stipule
53. d) Bud scale
54. b) Pinnate venation
55. a) Trichomes
56. d) Succulent leaf
57. b) Spongy mesophyll
58. b) Spine
59. a) Palmate venation
60. d) Parenchyma cells

INFLORESCENCE

1. What is the term used to describe the arrangement of flowers on the floral axis? a) Phyllotaxis b) Inflorescence c) Raceme d) Pedicel

2. Which of the following inflorescence types is characterized by a main axis with multiple lateral branches and flowers? a) Raceme b) Spike c) Umbel d) Corymb

3. In which inflorescence type are the flowers arranged along a single elongated axis, with the oldest flowers at the base and the youngest at the apex? a) Raceme b) Spike c) Panicle d) Spadix

4. Which inflorescence type is characterized by a flat-topped or convex shape, with the pedicels of the lower flowers being longer than those of the upper flowers? a) Umbel b) Corymb c) Cyme d) Head

5. What is the term used to describe an inflorescence in which the main axis terminates in a single flower? a) Umbel b) Spike c) Solitary d) Head

6. Which of the following inflorescence types is characterized by a dense, compact cluster of sessile flowers attached directly to a floral axis? a) Spike b) Head c) Raceme d) Spadix

7. What inflorescence type is often described as a flat-topped or rounded cluster of flowers in which the pedicels of all the flowers arise from a common point? a) Raceme b) Umbel c) Cyme d) Panicle

8. In which inflorescence type are the flowers arranged along a single unbranched axis, with the youngest flowers at the base and the oldest at the apex? a) Spike b) Raceme c) Panicle d) Spadix

9. What is the term used to describe an inflorescence in which the flowers are sessile and arranged in a more or less spherical cluster? a) Umbel b) Spike c) Head d) Cyme

10. Which inflorescence type is characterized by a main axis with multiple lateral branches, each ending in a single flower? a) Corymb b) Raceme c) Panicle d) Umbel

11. In which inflorescence type are the flowers arranged in a more or less flat-topped cluster, with the lower flowers having longer pedicels than the upper flowers? a) Raceme b) Corymb c) Spike d) Umbel

12. What is the term used to describe an inflorescence in which the flowers are arranged in a raceme-like manner, but the pedicels are of unequal length? a) Spadix b) Corymb c) Panicle d) Head

13. Which of the following inflorescence types is characterized by a thick, fleshy spike with tiny flowers embedded in the axis? a) Head b) Raceme c) Spadix d) Umbel

14. In which inflorescence type are the flowers arranged in a dense, rounded cluster, with the pedicels radiating from a common point like the spokes of an umbrella? a) Panicle b) Cyme c) Umbel d) Corymb

15. What is the term used to describe an inflorescence in which the flowers are arranged along a single elongated axis, with the pedicels of the lower flowers being longer than those of the upper flowers? a) Spike b) Panicle c) Raceme d) Spadix

16. Which inflorescence type is characterized by a long, unbranched axis with flowers attached directly to it, without any pedicels? a) Raceme b) Spike c) Corymb d) Head

17. In which inflorescence type are the flowers arranged in a flat-topped or convex cluster, with the pedicels of all the flowers arising from a common point? a) Cyme b) Umbel c) Raceme d) Panicle

18. What is the term used to describe an inflorescence in which the flowers are arranged in a more or less spherical cluster, with the youngest flowers at the center and the oldest at the periphery? a) Spike b) Cyme c) Head d) Raceme

19. Which of the following inflorescence types is characterized by a branched cluster of flowers, with each branch ending in a flower? a) Corymb b) Umbel c) Spike d) Panicle

20. In which inflorescence type are the flowers arranged in a more or less flat-topped cluster, with the pedicels of all the flowers arising from a common point? a) Raceme b) Cyme c) Umbel d) Head

21. What is the term used to describe an inflorescence in which the flowers are sessile and arranged in a more or less flat-topped cluster? a) Spike b) Head c) Cyme d) Umbel

22. Which inflorescence type is characterized by a main axis with multiple lateral branches, each ending in a cluster of flowers? a) Corymb b) Raceme c) Panicle d) Umbel

23. In which inflorescence type are the flowers arranged along a single elongated axis, with the oldest flowers at the apex and the youngest at the base? a) Raceme b) Spike c) Cyme d) Panicle

24. What is the term used to describe an inflorescence in which the flowers are attached directly to a floral axis without any pedicels? a) Umbel b) Spadix c) Raceme d) Head

25. Which of the following inflorescence types is characterized by a fleshy spike with small, inconspicuous flowers embedded in it? a) Head b) Raceme c) Spadix d) Umbel

26. In which inflorescence type are the flowers arranged in a dense, rounded cluster, with the pedicels of all the flowers attached to a common point? a) Panicle b) Cyme c) Umbel d) Corymb

27. What is the term used to describe an inflorescence in which the flowers are arranged along a single unbranched axis, with the pedicels of the lower flowers being longer than those of the upper flowers? a) Raceme b) Spike c) Panicle d) Spadix

28. Which inflorescence type is characterized by a long, unbranched axis with flowers attached directly to it, without any pedicels? a) Spike b) Raceme c) Corymb d) Head

29. In which inflorescence type are the flowers arranged in a flat-topped or convex cluster, with the pedicels of all the flowers arising from a common point? a) Cyme b) Umbel c) Raceme d) Panicle

30. What is the term used to describe an inflorescence in which the flowers are arranged in a more or less spherical cluster, with the youngest flowers at the center and the oldest at the periphery? a) Spike b) Cyme c) Head d) Raceme

31. Which of the following inflorescence types is characterized by a branched cluster of flowers, with each branch ending in a flower? a) Corymb b) Umbel c) Spike d) Panicle

32. In which inflorescence type are the flowers arranged in a more or less flat-topped cluster, with the pedicels of all the flowers arising from a common point? a) Raceme b) Cyme c) Umbel d) Head

33. What is the term used to describe an inflorescence in which the flowers are sessile and arranged in a more or less flat-topped cluster? a) Spike b) Head c) Cyme d) Umbel

34. Which inflorescence type is characterized by a main axis with multiple lateral branches, each ending in a cluster of flowers? a) Corymb b) Raceme c) Panicle d) Umbel

35. In which inflorescence type are the flowers arranged along a single elongated axis, with the oldest flowers at the apex and the youngest at the base? a) Raceme b) Spike c) Cyme d) Panicle

36. What is the term used to describe an inflorescence in which the flowers are attached directly to a floral axis without any pedicels? a) Umbel b) Spadix c) Raceme d) Head

37. Which of the following inflorescence types is characterized by a fleshy spike with small, inconspicuous flowers embedded in it? a) Head b) Raceme c) Spadix d) Umbel

38. In which inflorescence type are the flowers arranged in a dense, rounded cluster, with the pedicels of all the flowers attached to a common point? a) Panicle b) Cyme c) Umbel d) Corymb

39. What is the term used to describe an inflorescence in which the flowers are arranged along a single unbranched axis, with the pedicels of the lower flowers being longer than those of the upper flowers? a) Raceme b) Spike c) Panicle d) Spadix

40. Which inflorescence type is characterized by a long, unbranched axis with flowers attached directly to it, without any pedicels? a) Spike b) Raceme c) Corymb d) Head

41. In which inflorescence type are the flowers arranged in a flat-topped or convex cluster, with the pedicels of all the flowers arising from a common point? a) Cyme b) Umbel c) Raceme d) Panicle

42. What is the term used to describe an inflorescence in which the flowers are arranged in a more or less spherical cluster, with the youngest flowers at the center and the oldest at the periphery? a) Spike b) Cyme c) Head d) Raceme

43. Which of the following inflorescence types is characterized by a branched cluster of flowers, with each branch ending in a flower? a) Corymb b) Umbel c) Spike d) Panicle

44. In which inflorescence type are the flowers arranged in a more or less flat-topped cluster, with the pedicels of all the flowers arising from a common point? a) Raceme b) Cyme c) Umbel d) Head

45. What is the term used to describe an inflorescence in which the flowers are sessile and arranged in a more or less flat-topped cluster? a) Spike b) Head c) Cyme d) Umbel

46. Which inflorescence type is characterized by a main axis with multiple lateral branches, each ending in a cluster of flowers? a) Corymb b) Raceme c) Panicle d) Umbel

47. In which inflorescence type are the flowers arranged along a single elongated axis, with the oldest flowers at the apex and the youngest at the base? a) Raceme b) Spike c) Cyme d) Panicle

48. What is the term used to describe an inflorescence in which the flowers are attached directly to a floral axis without any pedicels? a) Umbel b) Spadix c) Raceme d) Head

49. Which of the following inflorescence types is characterized by a fleshy spike with small, inconspicuous flowers embedded in it? a) Head b) Raceme c) Spadix d) Umbel

50. In which inflorescence type are the flowers arranged in a dense, rounded cluster, with the pedicels of all the flowers attached to a common point? a) Panicle b) Cyme c) Umbel d) Corymb

51. What is the term used to describe an inflorescence in which the flowers are arranged along a single unbranched axis, with the pedicels of the lower flowers being longer than those of the upper flowers? a) Raceme b) Spike c) Panicle d) Spadix

52. Which inflorescence type is characterized by a long, unbranched axis with flowers attached directly to it, without any pedicels? a) Spike b) Raceme c) Corymb d) Head

53. In which inflorescence type are the flowers arranged in a flat-topped or convex cluster, with the pedicels of all the flowers arising from a common point? a) Cyme b) Umbel c) Raceme d) Panicle

54. What is the term used to describe an inflorescence in which the flowers are arranged in a more or less spherical cluster, with the youngest flowers at the center and the oldest at the periphery? a) Spike b) Cyme c) Head d) Raceme

55. Which of the following inflorescence types is characterized by a branched cluster of flowers, with each branch ending in a flower? a) Corymb b) Umbel c) Spike d) Panicle

56. In which inflorescence type are the flowers arranged in a more or less flat-topped cluster, with the pedicels of all the flowers arising from a common point? a) Raceme b) Cyme c) Umbel d) Head

57. What is the term used to describe an inflorescence in which the flowers are sessile and arranged in a more or less flat-topped cluster? a) Spike b) Head c) Cyme d) Umbel

58. Which inflorescence type is characterized by a main axis with multiple lateral branches, each ending in a cluster of flowers? a) Corymb b) Raceme c) Panicle d) Umbel

59. In which inflorescence type are the flowers arranged along a single elongated axis, with the oldest flowers at the apex and the youngest at the base? a) Raceme b) Spike c) Cyme d) Panicle

60. What is the term used to describe an inflorescence in which the flowers are attached directly to a floral axis without any pedicels? a) Umbel b) Spadix c) Raceme d) Head

61. Which of the following inflorescence types is characterized by a fleshy spike with small, inconspicuous flowers embedded in it? a) Head b) Raceme c) Spadix d) Umbel

62. In which inflorescence type are the flowers arranged in a dense, rounded cluster, with the pedicels of all the flowers attached to a common point? a) Panicle b) Cyme c) Umbel d) Corymb

63. What is the term used to describe an inflorescence in which the flowers are arranged along a single unbranched axis, with the pedicels of the lower flowers being longer than those of the upper flowers? a) Raceme b) Spike c) Panicle d) Spadix

64. Which inflorescence type is characterized by a long, unbranched axis with flowers attached directly to it, without any pedicels? a) Spike b) Raceme c) Corymb d) Head

65. In which inflorescence type are the flowers arranged in a flat-topped or convex cluster, with the pedicels of all the flowers arising from a common point? a) Cyme b) Umbel c) Raceme d) Panicle

66. What is the term used to describe an inflorescence in which the flowers are arranged in a more or less spherical cluster, with the youngest flowers at the center and the oldest at the periphery? a) Spike b) Cyme c) Head d) Raceme

67. Which of the following inflorescence types is characterized by a branched cluster of flowers, with each branch ending in a flower? a) Corymb b) Umbel c) Spike d) Panicle

68. In which inflorescence type are the flowers arranged in a more or less flat-topped cluster, with the pedicels of all the flowers arising from a common point? a) Raceme b) Cyme c) Umbel d) Head

69. What is the term used to describe an inflorescence in which the flowers are sessile and arranged in a more or less flat-topped cluster? a) Spike b) Head c) Cyme d) Umbel

70. Which inflorescence type is characterized by a main axis with multiple lateral branches, each ending in a cluster of flowers? a) Corymb b) Raceme c) Panicle d) Umbel

ANSWERS

1. b) Inflorescence
2. a) Raceme
3. b) Spike
4. b) Corymb
5. c) Solitary
6. d) Spadix
7. b) Umbel
8. a) Spike
9. c) Head
10. c) Panicle
11. b) Corymb
12. b) Corymb
13. c) Spadix
14. c) Umbel
15. c) Raceme
16. b) Spike
17. b) Umbel
18. c) Head
19. d) Panicle
20. c) Umbel
21. b) Head
22. a) Corymb
23. a) Raceme
24. b) Spadix
25. c) Spadix
26. c) Umbel
27. a) Raceme
28. a) Spike

29. b) Umbel

30. c) Head

31. a) Corymb

32. c) Umbel

33. b) Head

34. d) Umbel

35. a) Raceme

36. b) Spadix

37. c) Spadix

38. c) Umbel

39. a) Raceme

40. b) Spike

41. b) Umbel

42. c) Head

43. a) Corymb

44. c) Umbel

45. c) Cyme

46. a) Corymb

47. a) Raceme

48. b) Spadix

49. c) Spadix

50. c) Umbel

51. a) Raceme

52. b) Spike

53. b) Cyme

54. c) Head

55. a) Corymb

56. b) Cyme

57. d) Umbel

58. c) Panicle

59. a) Raceme

60. b) Spadix

61. c) Spadix

62. c) Umbel

63. a) Raceme

64. b) Spike

65. b) Umbel

66. c) Head

67. a) Corymb

68. b) Cyme

69. b) Head

70. c) Panicle

FLOWER

1. Which part of a flower produces pollen? a) Stamen b) Sepal c) Pistil d) Petal

2. What is the primary function of petals in a flower? a) To attract pollinators b) To produce nectar c) To protect the reproductive organs d) To support the stem

3. Which of the following is not a type of inflorescence? a) Raceme b) Spike c) Leaflet d) Umbel

4. The female reproductive organ in a flower is called the: a) Stamen b) Sepal c) Pistil d) Petal

5. What is the process by which pollen is transferred from the anther to the stigma of a flower? a) Fertilization b) Pollination c) Germination d) Photosynthesis

6. Which of the following is a type of incomplete flower? a) Sunflower b) Rose c) Orchid d) Daffodil

7. Which part of a flower develops into a fruit? a) Stamen b) Sepal c) Pistil d) Petal

8. What is the male reproductive part of a flower called? a) Stamen b) Sepal c) Pistil d) Petal

9. The process of a seed sprouting and developing into a new plant is known as: a) Germination b) Photosynthesis c) Fertilization d) Pollination

10. Which of the following is not a method of pollination? a) Wind b) Water c) Insects d) Photosynthesis

11. Which of the following flowers is commonly associated with Valentine's Day? a) Tulip b) Sunflower c) Lily d) Rose

12. What is the outermost part of a flower that protects the bud called? a) Stamen b) Sepal c) Pistil d) Petal

13. Which of the following is a type of asexual reproduction in plants? a) Pollination b) Fertilization c) Budding d) Germination

14. What is the process by which a plant produces its own food using sunlight, carbon dioxide, and water? a) Fertilization b) Pollination c) Germination d) Photosynthesis

15. Which part of a flower contains the ovules? a) Stamen b) Sepal c) Pistil d) Petal

16. Which of the following is a common example of an insect-pollinated flower? a) Dandelion b) Grass c) Orchid d) Willow tree

17. What is the function of the sepals in a flower? a) To attract pollinators b) To protect the flower bud c) To produce nectar d) To support the stem

18. Which of the following is a type of simple inflorescence? a) Raceme b) Spike c) Panicle d) Umbel

19. Which of the following is not a type of a modified stem? a) Rhizome b) Bulb c) Stolon d) Pistil

20. Which of the following is a method of seed dispersal in plants? a) Germination b) Photosynthesis c) Wind d) Fertilization

21. The brightly colored part of a flower that attracts pollinators is called the: a) Stamen b) Sepal c) Pistil d) Petal

22. Which of the following is a type of complete flower? a) Sunflower b) Rose c) Orchid d) Daffodil

23. What is the male part of a flower called, which consists of the anther and filament? a) Stamen b) Sepal c) Pistil d) Petal

24. Which of the following is an example of a wind-pollinated flower? a) Lily b) Sunflower c) Rose d) Oak tree

25. What is the process of transferring pollen from the anther of one flower to the stigma of another flower on a different plant called? a) Self-pollination b) Cross-pollination c) Asexual reproduction d) Vegetative propagation

26. Which part of a flower develops into a seed? a) Stamen b) Sepal c) Pistil d) Petal

27. What is the transfer of pollen grains from the anther to the stigma of the same flower called? a) Self-pollination b) Cross-pollination c) Asexual reproduction d) Vegetative propagation

28. Which of the following is a type of incomplete flower? a) Daisy b) Lily c) Orchid d) Marigold

29. The swollen base of the pistil that contains the ovules is called the: a) Anther b) Filament c) Stigma d) Ovary

30. What is the process of pollen grains germinating on the stigma called? a) Fertilization b) Pollination c) Germination d) Photosynthesis

31. Which of the following flowers is commonly associated with weddings? a) Tulip b) Sunflower c) Lily d) Rose

32. What is the small pore-like openings on the surface of leaves and stems that allow gases and water vapor to pass through called? a) Stomata b) Chloroplasts c) Xylem d) Phloem

33. The process by which water is lost from a plant through the leaves is called: a) Transpiration b) Respiration c) Osmosis d) Photosynthesis

34. Which of the following is a type of a modified leaf? a) Stolon b) Thorn c) Tendril d) Pistil

35. What is the primary function of the sepals in a flower? a) To attract pollinators b) To protect the reproductive organs c) To produce nectar d) To support the stem

36. Which of the following is a type of specialized underground stem? a) Rhizome b) Bulb c) Stolon d) Pistil

37. The stalk that supports the anther in a flower is called the: a) Anther b) Filament c) Stigma d) Ovary

38. Which of the following is a method of asexual reproduction in plants? a) Pollination b) Fertilization c) Budding d) Germination

39. What is the process by which plants bend or grow towards a source of light called? a) Phototropism b) Gravitropism c) Hydrotropism d) Thigmotropism

40. Which of the following is a common example of a bird-pollinated flower? a) Dandelion b) Grass c) Orchid d) Lily

41. What is the process of transferring pollen from the anther to the stigma of the same flower called? a) Self-pollination b) Cross-pollination c) Asexual reproduction d) Vegetative propagation

42. Which of the following is a type of complete flower? a) Daisy b) Lily c) Orchid d) Marigold

43. What is the transfer of pollen grains from the anther to the stigma of another flower on the same plant called? a) Self-pollination b) Cross-pollination c) Asexual reproduction d) Vegetative propagation

44. Which part of a flower produces eggs? a) Stamen b) Sepal c) Pistil d) Petal

45. What is the process by which a seed becomes a mature plant called? a) Germination b) Photosynthesis c) Fertilization d) Pollination

46. Which of the following is a common example of a wind-dispersed fruit? a) Apple b) Coconut c) Strawberry d) Orange

47. What is the primary purpose of the nectar produced by flowers? a) To attract pollinators b) To store energy c) To protect the reproductive organs d) To support the stem

48. Which of the following is a type of specialized aboveground stem? a) Rhizome b) Bulb c) Stolon d) Tendril

49. What is the male reproductive part of a flower called? a) Stamen b) Sepal c) Pistil d) Petal

50. What is the process by which water is absorbed by plant roots called? a) Transpiration b) Respiration c) Osmosis d) Absorption

51. Which of the following is an example of an epiphytic plant that grows on other plants or structures? a) Rose b) Orchid c) Sunflower d) Tulip

52. What is the process of transferring pollen from the anther to the stigma of a flower of the same plant called? a) Autogamy b) Dichogamy c) Geitonogamy d) Allogamy

53. Which of the following flower structures is responsible for producing pollen? a) Style b) Ovule c) Anther d) Receptacle

54. What is the term used to describe the fusion of male and female gametes in a flower? a) Fertilization b) Pollination c) Germination d) Meiosis

55. Which of the following is a type of flower arrangement where the flowers are arranged in a flat-topped cluster? a) Raceme b) Spike c) Corymb d) Capitulum

56. Which of the following is a flower color that attracts bees and other insect pollinators? a) Red b) Blue c) White d) Green

57. What is the term used to describe a flower that has both male and female reproductive organs? a) Monoecious b) Dioecious c) Hermaphroditic d) Bisexual

58. Which of the following is a plant hormone responsible for promoting flower bud formation? a) Auxin b) Cytokinin c) Gibberellin d) Ethylene

59. What is the process of transferring pollen from the anther to the stigma of a flower of a different plant called? a) Autogamy b) Dichogamy c) Geitonogamy d) Allogamy

60. Which of the following flower structures develops into the fruit? a) Stamen b) Sepal c) Ovary d) Petal

61. Which of the following flower shapes is characterized by a long, tubular corolla? a) Funnel-shaped b) Bell-shaped c) Star-shaped d) Trumpet-shaped

62. What is the term used to describe the shedding of leaves or flowers in response to environmental conditions? a) Senescence b) Abcission c) Dormancy d) Etiolation

63. Which of the following is a flower adaptation that attracts pollinators through scent? a) Nectar guides b) Spur c) Bracts d) Perfumes

64. What is the process of a plant producing new flowers and fruits after the initial ones have withered called? a) Rejuvenation b) Regeneration c) Reforestation d) Reblooming

65. Which of the following is a flower shape characterized by a circular arrangement of petals? a) Radial b) Bilateral c) Rotate d) Composite

66. What is the term used to describe a flower that lacks one or more of the typical floral parts? a) Incomplete b) Imperfect c) Unisexual d) Perfect

67. Which of the following is a type of flower pollinated by birds? a) Bee orchid b) Hummingbird sage c) Bee balm d) Butterfly weed

ANSWERS

1.   a) Stamen

2.   a) To attract pollinators

3.   c) Leaflet

4.   c) Pistil

5.   b) Pollination

6.   d) Daffodil

7.   c) Pistil

8.   a) Stamen

9.   a) Germination

10. d) Photosynthesis

11. d) Rose

12. b) Sepal

13. c) Budding

14. d) Photosynthesis

15. c) Pistil

16. c) Orchid

17. b) To protect the flower bud

18. a) Raceme

19. d) Pistil

20. c) Wind

21. d) Petal

22. b) Rose

23. a) Stamen

24. d) Oak tree

25. b) Cross-pollination

26. d) Petal

27. a) Self-pollination

28. c) Orchid

29. d) Ovary

30. b) Pollination

31. d) Rose

32. a) Stomata

33. a) Transpiration

34. b) Thorn

35. b) To protect the reproductive organs

36. a) Rhizome

37. b) Filament

38. c) Budding

39. a) Phototropism

40. d) Lily

41. a) Self-pollination

42. b) Lily

43. b) Cross-pollination

44. c) Pistil

45. a) Germination

46. b) Coconut

47. a) To attract pollinators

48. d) Tendril

49. a) Stamen

50. d) Absorption

51. b) Orchid

52. a) Autogamy

53. c) Anther

54. a) Fertilization

55. c) Corymb

56. a) Red

57. c) Hermaphroditic

58. c) Gibberellin

59. d) Allogamy

60. c) Ovary

61. d) Trumpet-shaped

62. b) Abcission

63. d) Perfumes

64. d) Reblooming

65. a) Radial

66. a) Incomplete

67. b) Hummingbird sage

CHAPTER 2: PLANT PHYSIOLOGY

PHOTOSYNTHESIS

1. What is the primary pigment responsible for capturing light energy in photosynthesis?
a) Chlorophyll
b) Carotenoid
c) Xanthophyll
d) Phycobilin

2. In which organelle does photosynthesis occur in plant cells?
a) Chloroplast
b) Mitochondria
c) Nucleus
d) Vacuole

3. Which of the following is NOT a product of photosynthesis?
a) Oxygen
b) Glucose
c) Carbon dioxide
d) Water

4. What is the ultimate source of energy for photosynthesis?
a) Sunlight
b) Water
c) Carbon dioxide
d) Glucose

5. During photosynthesis, light energy is converted into chemical energy stored in the form of:
a) ATP
b) ADP
c) NADPH
d) FADH2

6. What is the primary function of the light-dependent reactions in photosynthesis?
a) Capture light energy and convert it into chemical energy
b) Convert carbon dioxide into glucose
c) Produce oxygen as a by-product
d) Generate ATP and NADPH

7. Which of the following is NOT a step of the light-dependent reactions in photosynthesis?
a) Photosystem II
b) Electron transport chain
c) Calvin cycle
d) Photosystem I

8. What is the primary function of the Calvin cycle in photosynthesis?
a) Fix carbon dioxide and produce glucose
b) Capture light energy and convert it into chemical energy
c) Generate ATP and NADPH
d) Produce oxygen as a by-product

9. During the Calvin cycle, carbon dioxide is converted into glucose through a series of chemical reactions that require:
a) ATP and NADPH
b) Sunlight
c) Oxygen
d) Water

10. What is the role of water in the light-dependent reactions of photosynthesis?
a) It donates electrons to photosystem II
b) It serves as a source of carbon dioxide
c) It generates ATP and NADPH
d) It produces glucose as a by-product

11. Which of the following colors of light is most effective in driving photosynthesis?
a) Red
b) Green
c) Blue
d) Yellow

12. What is the term for the process by which plants open and close small pores called stomata to regulate gas exchange during photosynthesis?
a) Transpiration
b) Respiration
c) Photolysis
d) Translocation

13. In which part of the chloroplast do the light-dependent reactions occur?
a) Thylakoid membrane
b) Stroma

c) Outer membrane
d) Inner membrane

14. What is the primary function of the electron transport chain in photosynthesis?
a) Generate ATP and NADPH
b) Convert carbon dioxide into glucose
c) Produce oxygen as a by-product
d) Capture light energy and convert it into chemical energy

15. Which molecule acts as the final electron acceptor in the light-dependent reactions of photosynthesis?
a) NADP+
b) ATP
c) Oxygen
d) Carbon dioxide

16. What is the primary function of the pigment molecules in the photosystems of the chloroplasts?
a) Absorb light energy
b) Store glucose
c) Produce oxygen
d) Convert ATP into ADP

17. What is the equation for photosynthesis?
a) $6CO_2 + 6H_2O \rightarrow C_6H_{12}O_6 + 6O_2$
b) $C_6H_{12}O_6 + 6O_2 \rightarrow 6CO_2 + 6H_2O$
c) $6CO_2 + 6O_2 \rightarrow C_6H_{12}O_6 + 6H_2O$
d) $C_6H_{12}O_6 + 6H_2O \rightarrow 6CO_2 + 6O_2$

18. What is the role of ATP synthase in photosynthesis?
a) It generates ATP from ADP and inorganic phosphate
b) It converts glucose into ATP
c) It absorbs light energy and transfers it to the chlorophyll molecules
d) It produces oxygen as a by-product

19. What is the primary function of the enzyme RuBisCO in photosynthesis?
a) Fix carbon dioxide during the Calvin cycle
b) Generate ATP and NADPH
c) Produce glucose as a by-product
d) Absorb light energy

20. What is the purpose of the oxygen produced during photosynthesis?
a) It is released into the atmosphere as a by-product
b) It is used in the Calvin cycle to produce glucose
c) It serves as a source of energy for the plant
d) It is converted into water through photolysis

21. What is the term for the process by which plants convert glucose into energy in the absence of light?
a) Cellular respiration
b) Photosynthesis
c) Fermentation
d) Transpiration

22. What is the role of NADPH in photosynthesis?
a) It carries high-energy electrons to the Calvin cycle
b) It donates electrons to photosystem I
c) It generates ATP through chemiosmosis
d) It absorbs light energy and transfers it to chlorophyll

23. Which environmental factor can directly affect the rate of photosynthesis?
a) Light intensity
b) Temperature
c) Carbon dioxide concentration
d) All of the above

24. What is the primary function of the stomata in photosynthesis?
a) Regulate gas exchange by opening and closing
b) Absorb light energy
c) Store glucose
d) Produce oxygen as a by-product

25. How do C4 plants minimize the effects of photo-respiration?
a) By spatially separating the initial carbon dioxide fixation from the Calvin cycle
b) By increasing the concentration of oxygen in the chloroplasts
c) By reducing the efficiency of the electron transport chain
d) By decreasing the concentration of carbon dioxide in the leaves

26. Which of the following is a product of the light-dependent reactions in photosynthesis?
a) ATP
b) Carbon dioxide
c) Glucose
d) RuBP

27. What is the primary function of the antenna pigments in photosynthesis?
a) Absorb light energy and transfer it to the reaction center chlorophyll
b) Fix carbon dioxide during the Calvin cycle
c) Produce ATP and NADPH
d) Generate oxygen as a by-product

28. Which wavelengths of light are most effective in driving photosynthesis?
a) Red and blue
b) Green and yellow
c) Violet and orange
d) Infrared and ultraviolet

29. What is the role of the proton gradient in the thylakoid membrane during photosynthesis?
a) It generates ATP through chemiosmosis
b) It transports electrons in the electron transport chain
c) It regulates the opening and closing of stomata
d) It converts glucose into carbon dioxide

30. What is the primary function of the carotenoid pigments in photosynthesis?
a) Protect chlorophyll from damage by excess light energy
b) Absorb carbon dioxide from the atmosphere
c) Produce glucose as a by-product
d) Release oxygen into the atmosphere

31. Which of the following is NOT a factor that can limit the rate of photosynthesis?
a) Water availability
b) Oxygen concentration
c) Temperature
d) Carbon dioxide concentration

32. How does light intensity affect the rate of photosynthesis?
a) Higher light intensity increases the rate of photosynthesis until it reaches a maximum level
b) Higher light intensity decreases the rate of photosynthesis
c) Light intensity has no effect on the rate of photosynthesis
d) Higher light intensity inhibits the process of photosynthesis

33. What is the function of the electron carriers NADP+ and NADPH in photosynthesis?
a) They shuttle high-energy electrons between the light-dependent reactions and the Calvin cycle
b) They absorb light energy and transfer it to the chlorophyll molecules

c) They convert glucose into ATP and NADPH
d) They produce oxygen as a by-product

34. How does temperature affect the rate of photosynthesis?
a) Higher temperatures increase the rate of photosynthesis until it reaches an optimal temperature, beyond which it decreases
b) Higher temperatures always increase the rate of photosynthesis
c) Temperature has no effect on the rate of photosynthesis
d) Higher temperatures inhibit the process of photosynthesis

35. What is the primary function of the ATP produced during photosynthesis?
a) It provides energy for cellular processes within the plant
b) It donates high-energy electrons to the Calvin cycle
c) It absorbs light energy and transfers it to the reaction center chlorophyll
d) It converts glucose into carbon dioxide

36. What is the primary function of the NADPH produced during photosynthesis?
a) It provides reducing power for the Calvin cycle
b) It releases oxygen into the atmosphere
c) It generates ATP through chemiosmosis
d) It absorbs light energy and transfers it to the chlorophyll molecules

37. What is the role of the enzyme phosphoenolpyruvate carboxylase (PEPCase) in C4 plants?
a) It facilitates the initial carbon dioxide fixation in mesophyll cells
b) It regulates the opening and closing of stomata
c) It produces glucose as a by-product
d) It generates ATP through substrate-level phosphorylation

38. Which of the following is an adaptation of CAM plants to reduce water loss during photosynthesis?
a) Opening stomata only at night
b) Increasing the concentration of carbon dioxide in the leaves
c) Fixing carbon dioxide during the Calvin cycle
d) Absorbing light energy and transferring it to the reaction center chlorophyll

39. How do C3 plants minimize water loss during photosynthesis?
a) By regulating the opening and closing of stomata
b) By spatially separating the initial carbon dioxide fixation from the Calvin cycle
c) By increasing the concentration of carbon dioxide in the chloroplasts
d) By reducing the efficiency of the electron transport chain

40. Which of the following is a by-product of the Calvin cycle in photosynthesis?
a) RuBP
b) Oxygen
c) ATP
d) NADPH

41. What is the primary function of the enzyme ribulose-1,5-bisphosphate carboxylase/oxygenase (RuBisCO) in the Calvin cycle?
a) Fix carbon dioxide during the carbon fixation step
b) Generate ATP and NADPH
c) Produce glucose as a by-product
d) Absorb light energy

42. Which of the following factors would most likely limit the rate of photosynthesis in aquatic plants?
a) Availability of carbon dioxide
b) Light intensity
c) Oxygen concentration
d) Temperature

43. How do CAM plants minimize water loss during photosynthesis?
a) By opening stomata only at night
b) By increasing the concentration of oxygen in the chloroplasts
c) By reducing the efficiency of the electron transport chain
d) By decreasing the concentration of carbon dioxide in the leaves

44. What is the function of the enzyme ATP synthase in photosynthesis?
a) It generates ATP from ADP and inorganic phosphate
b) It converts glucose into ATP
c) It absorbs light energy and transfers it to the chlorophyll molecules
d) It produces oxygen as a by-product

45. How does carbon dioxide concentration affect the rate of photosynthesis?
a) Higher carbon dioxide concentration increases the rate of photosynthesis until it reaches a maximum level
b) Higher carbon dioxide concentration decreases the rate of photosynthesis
c) Carbon dioxide concentration has no effect on the rate of photosynthesis
d) Higher carbon dioxide concentration inhibits the process of photosynthesis

46. What is the role of photolysis in the light-dependent reactions of photosynthesis?
a) Split water molecules and release oxygen into the atmosphere
b) Absorb light energy and transfer it to the reaction center chlorophyll

c) Generate ATP and NADPH
d) Fix carbon dioxide during the Calvin cycle

47. How does light intensity affect the opening and closing of stomata during photosynthesis?
a) Higher light intensity causes stomata to open wider
b) Higher light intensity causes stomata to close completely
c) Light intensity has no effect on the opening and closing of stomata
d) Light intensity affects the opening and closing of stomata indirectly through temperature regulation

48. What is the purpose of the cyclic electron flow during photosynthesis?
a) Generate additional ATP
b) Produce glucose as a by-product
c) Release oxygen into the atmosphere
d) Convert carbon dioxide into carbohydrates

49. What is the primary function of the enzyme phosphoglycerate kinase in the Calvin cycle?
a) Generate ATP from ADP and inorganic phosphate
b) Fix carbon dioxide during the carbon fixation step
c) Produce glucose as a by-product
d) Absorb light energy

50. How does the rate of photosynthesis vary with increasing temperature?
a) Initially increases, then decreases
b) Remains constant
c) Increases linearly
d) Decreases linearly

51. Which of the following is a strategy used by desert plants to minimize water loss during photosynthesis?
a) CAM photosynthesis
b) C4 photosynthesis
c) C3 photosynthesis
d) Photosynthesis in aquatic plants

ANSWER

1. a) Chlorophyll
2. a) Chloroplast
3. c) Carbon dioxide
4. a) Sunlight
5. a) ATP
6. a) Capture light energy and convert it into chemical energy
7. c) Calvin cycle
8. a) Fix carbon dioxide and produce glucose
9. a) ATP and NADPH
10. a) It donates electrons to photosystem II
11. a) Red
12. a) Transpiration
13. a) Thylakoid membrane
14. a) Generate ATP and NADPH
15. a) NADP+
16. a) Absorb light energy
17. a) $6CO_2 + 6H_2O \rightarrow C_6H_{12}O_6 + 6O_2$
18. a) It generates ATP from ADP and inorganic phosphate
19. a) Fix carbon dioxide during the Calvin cycle
20. a) It is released into the atmosphere as a by-product
21. c) Fermentation
22. a) It carries high-energy electrons to the Calvin cycle
23. d) All of the above
24. a) Regulate gas exchange by opening and closing
25. a) By spatially separating the initial carbon dioxide fixation from the Calvin cycle
26. a) ATP
27. a) Absorb light energy and transfer it to the reaction center chlorophyll
28. a) Red and blue

29. a) It generates ATP through chemiosmosis

30. a) Protect chlorophyll from damage by excess light energy

31. b) Oxygen concentration

32. a) Higher light intensity increases the rate of photosynthesis until it reaches a maximum level

33. a) They shuttle high-energy electrons between the light-dependent reactions and the Calvin cycle

34. a) Higher temperatures increase the rate of photosynthesis until it reaches an optimal temperature, beyond which it decreases

35. a) It provides energy for cellular processes within the plant

36. a) It provides reducing power for the Calvin cycle

37. a) Facilitate the initial carbon dioxide fixation in mesophyll cells

38. a) Opening stomata only at night

39. a) By regulating the opening and closing of stomata

40. d) NADPH

41. a) Fix carbon dioxide during the carbon fixation step

42. a) Availability of carbon dioxide

43. a) By opening stomata only at night

44. a) It generates ATP from ADP and inorganic phosphate

45. a) Higher carbon dioxide concentration increases the rate of photosynthesis until it reaches a maximum level

46. a) Split water molecules and release oxygen into the atmosphere

47. a) Higher light intensity causes stomata to open wider

48. a) Generate additional ATP

49. a) Generate ATP from ADP and inorganic phosphate

50. a) Initially increases, then decreases

51. a) CAM photosynthesis

RESPIRATION

1. Which of the following is the primary process by which plants obtain oxygen for respiration? a) Photosynthesis b) Transpiration c) Diffusion d) Respiration

2. What is the main product of plant respiration? a) Oxygen b) Carbon dioxide c) Glucose d) Water

3. In which part of the plant does cellular respiration predominantly occur? a) Leaves b) Stems c) Roots d) Flowers

4. Which of the following is NOT a requirement for plant respiration? a) Oxygen b) Carbon dioxide c) Water d) Light

5. Which cellular organelle is responsible for the majority of plant respiration? a) Nucleus b) Mitochondria c) Chloroplasts d) Endoplasmic reticulum

6. What is the primary purpose of plant respiration? a) Energy production b) Carbon dioxide absorption c) Water uptake d) Photosynthesis

7. During respiration, plants release energy stored in which molecule? a) Glucose b) Starch c) Cellulose d) Sucrose

8. Which of the following is a byproduct of plant respiration? a) Oxygen b) Carbon dioxide c) Glucose d) Chlorophyll

9. Where does the initial step of plant respiration occur? a) Cytoplasm b) Nucleus c) Mitochondria d) Cell membrane

10. Which of the following is an essential component for the process of aerobic respiration in plants? a) Sunlight b) Nitrogen c) Oxygen d) Phosphorus

11. In plant respiration, which molecule is broken down to release energy? a) Starch b) Cellulose c) Sucrose d) Glucose

12. Which type of respiration occurs in the absence of oxygen in plants? a) Aerobic respiration b) Anaerobic respiration c) Photorespiration d) Transpiration

13. What is the primary waste product of anaerobic respiration in plants? a) Oxygen b) Carbon dioxide c) Ethanol d) Water

14. During aerobic respiration, what is the final electron acceptor in the electron transport chain? a) Carbon dioxide b) Oxygen c) Glucose d) ATP

15. In plants, which process directly produces ATP during respiration? a) Glycolysis b) Krebs cycle c) Electron transport chain d) Fermentation

16. Which of the following is an example of an environmental factor that can affect plant respiration? a) Light intensity b) Soil pH c) Temperature d) Wind speed

17. What is the primary purpose of respiration in plant roots? a) Absorption of water b) Absorption of carbon dioxide c) Absorption of nutrients d) Production of glucose

18. During respiration, what happens to the glucose molecules in plants?
a) They are broken down to release energy.
b) They are converted to starch for storage.
c) They are transported to the leaves for photosynthesis.
d) They are converted to cellulose for cell wall synthesis.

19. Which of the following is NOT a stage of aerobic respiration in plants? a) Glycolysis b) Krebs cycle c) Calvin cycle d) Electron transport chain

20. Which part of the plant is primarily responsible for the intake of oxygen during respiration? a) Leaves b) Stems c) Roots d) Flowers

21. What is the net ATP yield from one molecule of glucose during aerobic respiration in plants? a) 2 ATP b) 10 ATP c) 38 ATP d) 60 ATP

22. In plant respiration, what is the purpose of the Krebs cycle? a) Production of ATP b) Generation of carbon dioxide c) Breakdown of glucose molecules d) Transfer of electrons

23. What is the term used to describe the overall process of respiration in plants? a) Glycolysis b) Fermentation c) Oxidative phosphorylation d) Cellular respiration

24. During respiration, what gas is exchanged between plant cells and the external environment? a) Oxygen b) Carbon dioxide c) Nitrogen d) Hydrogen

25. Which of the following is a product of anaerobic respiration in plants? a) Oxygen b) Carbon dioxide c) Ethanol d) Glucose

26. In which type of plant tissue does respiration occur most actively? a) Epidermis b) Xylem c) Phloem d) Parenchyma

27. What is the primary metabolic pathway involved in the breakdown of glucose during respiration in plants? a) Calvin cycle b) Glycolysis c) Krebs cycle d) Photosynthesis

28. Which of the following is an example of an environmental condition that can inhibit plant respiration? a) High humidity b) Low light intensity c) Acidic soil d) Excessive rainfall

29. How do plant cells obtain oxygen for respiration? a) Through stomata b) Through cell walls c) Through the cuticle d) Through the vacuole

30. Which of the following is a characteristic of anaerobic respiration in plants?
a) It produces more ATP than aerobic respiration.
b) It requires the presence of oxygen.
c) It occurs in the mitochondria.
d) It results in the production of lactic acid.

31. What is the primary function of respiration in plants? a) Energy production b) Oxygen production c) Water absorption d) Nutrient transport

32. Which of the following is an example of an environmental factor that can increase plant respiration rate? a) High humidity b) Low temperature c) Excessive light intensity d) Alkaline soil

33. What is the term used to describe the process of converting glucose to energy in plant cells? a) Photosynthesis b) Respiration c) Transpiration d) Germination

34. What is the primary source of glucose for plant respiration? a) Carbon dioxide b) Water c) Sunlight d) Organic compounds

35. Which of the following is an example of an environmental factor that can decrease plant respiration rate? a) High temperature b) Low light intensity c) Acidic soil d) Excessive rainfall

36. What is the purpose of the electron transport chain in plant respiration? a) Production of ATP b) Generation of carbon dioxide c) Breakdown of glucose molecules d) Transfer of electrons

37. In plant respiration, what is the final electron acceptor in the electron transport chain? a) Carbon dioxide b) Oxygen c) Glucose d) ATP

38. Which of the following is a byproduct of aerobic respiration in plants? a) Oxygen b) Carbon dioxide c) Ethanol d) Water

39. How does respiration differ from photosynthesis in plants?
a) Respiration occurs in the presence of light, while photosynthesis occurs in darkness.
b) Respiration releases energy, while photosynthesis requires energy.
c) Respiration produces oxygen, while photosynthesis produces carbon dioxide.
d) Respiration occurs in the roots, while photosynthesis occurs in the leaves.

40. Which process involves the breakdown of glucose in the absence of oxygen in plants?
a) Aerobic respiration b) Anaerobic respiration c) Photorespiration d) Transpiration

ANSWERS

1. c) Diffusion
2. b) Carbon dioxide
3. b) Stems
4. d) Light
5. b) Mitochondria
6. a) Energy production
7. a) Oxygen
8. b) Carbon dioxide
9. c) Mitochondria
10. c) Oxygen
11. a) Glucose
12. b) Anaerobic respiration
13. c) Ethanol
14. b) Oxygen
15. c) Electron transport chain
16. c) Temperature
17. a) Absorption of water
18. a) They are broken down to release energy.
19. c) Calvin cycle
20. c) Roots
21. c) 38 ATP
22. b) Generation of carbon dioxide
23. d) Cellular respiration
24. b) Carbon dioxide
25. c) Ethanol
26. d) Parenchyma
27. b) Glycolysis
28. b) Low light intensity
29. a) Through stomata

30. d) It results in the production of lactic acid.

31. a) Energy production

32. c) Excessive light intensity

33. b) Respiration

34. d) Organic compounds

35. b) Low light intensity

36. a) Production of ATP

37. b) Oxygen

38. d) Water

39. b) Respiration releases energy, while photosynthesis requires energy.

40. b) Anaerobic respiration

TRANSPIRATION

1. What is plant transpiration?
a) The process of water uptake by plant roots
b) The process of sugar transport within plants
c) The process of water loss through plant leaves
d) The process of carbon dioxide absorption by plants

2. Which part of the plant is primarily responsible for transpiration? a) Leaves b) Stems c) Roots d) Flowers

3. What drives the process of transpiration in plants? a) Photosynthesis b) Respiration c) Osmosis d) Evaporation

4. Which of the following is a factor that affects the rate of transpiration in plants? a) Light intensity b) Soil pH c) Carbon dioxide concentration d) Wind direction

5. What is the primary function of transpiration in plants? a) Nutrient absorption b) Water absorption c) Cooling of the plant d) Sugar production

6. Which structure in plants regulates the rate of transpiration? a) Guard cells b) Xylem vessels c) Phloem sieve tubes d) Stomata

7. What is the term used to describe the loss of water vapor from the stomata of plant leaves? a) Respiration b) Osmosis c) Transpiration d) Photosynthesis

8. How does transpiration contribute to the movement of water within plants? a) Through the xylem b) Through the phloem c) Through the stomata d) Through the roots

9. Which environmental factor is most closely associated with an increase in transpiration rate? a) High humidity b) Low temperature c) Still air d) Bright sunlight

10. Which of the following is NOT a function of transpiration in plants? a) Nutrient absorption b) Cooling of the plant c) Transport of minerals d) Maintenance of cell turgidity

11. What is the process by which water is transported from the roots to the leaves of a plant? a) Osmosis b) Photosynthesis c) Transpiration pull d) Capillary action

12. Which type of transpiration occurs through the stomata on the leaf surface?
a) Cuticular transpiration
b) Lenticular transpiration

c) Stomatal transpiration
d) Subterranean transpiration

13. What is the term used to describe the opening and closing of stomata to regulate transpiration? a) Osmosis b) Translocation c) Tropism d) Guard cell movement

14. How does wind affect the rate of transpiration in plants?
a) It decreases transpiration rate
b) It increases transpiration rate
c) It has no effect on transpiration rate
d) It depends on the plant species

15. Which environmental factor is most closely associated with a decrease in transpiration rate? a) Low humidity b) High temperature c) Windy conditions d) Bright sunlight

16. Which of the following is a product of transpiration in plants? a) Oxygen b) Carbon dioxide c) Water vapor d) Glucose

17. How does transpiration contribute to the cooling of plants? a) By absorbing heat from the soil b) By releasing water vapor from the leaves c) By converting sunlight into energy d) By facilitating the process of respiration

18. Which part of the plant absorbs water from the soil for transpiration? a) Roots b) Leaves c) Flowers d) Stems

19. What is the term used to describe the movement of water from the roots to the leaves of a plant? a) Transpiration b) Osmosis c) Photosynthesis d) Capillary action

20. Which of the following environmental factors does NOT affect transpiration rate? a) Light intensity b) Humidity c) Wind speed d) Soil pH

21. What is the primary driving force behind transpiration? a) Photosynthesis b) Osmosis c) Capillary action d) Evaporation

22. Which of the following is a consequence of excessive transpiration in plants? a) Wilting b) Leaf discoloration c) Reduced nutrient uptake d) Enhanced growth

23. Which part of the plant allows gases to enter and exit during transpiration? a) Guard cells b) Xylem vessels c) Phloem sieve tubes d) Stomata

24. Which type of transpiration occurs through the surface of the stem and other above-ground plant parts?

a) Cuticular transpiration
b) Lenticular transpiration
c) Stomatal transpiration
d) Subterranean transpiration

25. How does light intensity affect the rate of transpiration in plants?
a) It increases transpiration rate
b) It decreases transpiration rate
c) It has no effect on transpiration rate
d) It depends on the plant species

26. Which part of the plant stores excess water to reduce transpiration rate? a) Leaves b) Stems c) Roots d) Flowers

27. What is the process by which plants lose water vapor through their cuticle called?
a) Cuticular transpiration
b) Lenticular transpiration
c) Stomatal transpiration
d) Subterranean transpiration

28. How does humidity affect the rate of transpiration in plants?
a) High humidity decreases transpiration rate
b) High humidity increases transpiration rate
c) Low humidity decreases transpiration rate
d) Low humidity increases transpiration rate

29. Which part of the plant has specialized structures called hydathodes that release water droplets during transpiration? a) Leaves b) Stems c) Roots d) Flowers

30. What is the term used to describe the force that pulls water up from the roots to the leaves during transpiration? a) Cohesion b) Adhesion c) Turgor pressure d) Transpiration pull

31. How does temperature affect the rate of transpiration in plants?
a) High temperature increases transpiration rate
b) High temperature decreases transpiration rate
c) Low temperature increases transpiration rate
d) Low temperature decreases transpiration rate

32. Which of the following factors can cause wilting in plants due to excessive transpiration? a) Low light intensity b) High humidity c) High soil moisture d) Strong wind

33. What is the term used to describe the process by which water evaporates from the surface of plant leaves during transpiration? a) Respiration b) Osmosis c) Diffusion d) Evapo-transpiration

34. How does the size and density of stomata affect transpiration rate in plants?
a) Larger stomata increase transpiration rate
b) Larger stomata decrease transpiration rate
c) Higher stomatal density increases transpiration rate
d) Higher stomatal density decreases transpiration rate

35. Which of the following environmental factors can lead to an increase in transpiration rate? a) Low light intensity b) High humidity c) Low wind speed d) Low temperature

36. What is the role of the Casparian strip in plant roots in relation to transpiration?
a) It regulates the opening and closing of stomata
b) It prevents the back-flow of water in the xylem
c) It enhances the absorption of carbon dioxide
d) It facilitates the movement of sugars in the phloem

37. Which of the following is an example of a plant adaptation to reduce excessive transpiration? a) Thick cuticle b) Large stomatal aperture c) Shallow root system d) Enhanced shoot elongation

38. Which of the following structures is responsible for the movement of water within the xylem during transpiration? a) Tracheids b) Sieve tubes c) Companion cells d) Phloem fibers

39. How does transpiration in plants contribute to the absorption of nutrients from the soil?
a) It creates a concentration gradient for nutrient uptake
b) It directly transports nutrients from the roots to the leaves
c) It stimulates the release of nutrients from micro-organisms in the soil
d) It increases the availability of nutrients in the soil solution

40. Which environmental factor has the greatest influence on the rate of transpiration in plants? a) Light intensity b) Temperature c) Humidity d) Wind speed

41. What is the term used to describe the process by which water moves from the xylem into the mesophyll cells during transpiration?
a) Root pressure
b) Capillary action
c) Osmosis

d) Cell turgor

42. How does transpiration in plants contribute to the uptake and transport of minerals from the soil?
a) It creates a pressure gradient for mineral absorption
b) It directly transports minerals from the roots to the leaves
c) It stimulates the release of minerals from the root hairs
d) It increases the solubility of minerals in the soil solution

43. Which of the following plant hormones can regulate the opening and closing of stomata to control transpiration rate? a) Auxin b) Cytokinin c) Gibberellin d) Abscisic acid

44. How does the structure of the root system in plants affect transpiration rate?
a) A deep root system increases transpiration rate
b) A shallow root system increases transpiration rate
c) A dense root system decreases transpiration rate
d) A sparse root system decreases transpiration rate

45. What is the term used to describe the process by which water moves from the soil into the roots during transpiration?
a) Root pressure
b) Capillary action
c) Osmosis
d) Cell turgor

46. How does the presence of air pollutants, such as sulfur dioxide, affect transpiration in plants?
a) It decreases transpiration rate
b) It increases transpiration rate
c) It has no effect on transpiration rate
d) It depends on the plant species

47. Which of the following environmental conditions can lead to an increase in transpiration rate?
a) Low light intensity and high humidity
b) High temperature and low wind speed
c) High humidity and low temperature
d) High wind speed and low light intensity

48. What is the term used to describe the process by which water vapor is released from the plant surface into the surrounding air during transpiration?

a) Evaporation
b) Condensation
c) c) Sublimation
d) d) Precipitation

49. How does transpiration contribute to the movement of nutrients within plants?
a) By actively transporting nutrients in the xylem
b) By creating a pressure gradient for nutrient uptake
c) By facilitating the diffusion of nutrients through the stomata
d) By increasing the solubility of nutrients in the soil solution

50. Which of the following is a method used by plants to reduce excessive transpiration?
a) Closing stomata during the day
b) Opening stomata during the night
c) Increasing leaf surface area
d) Decreasing root length

51. What is the term used to describe the process by which water moves from cell to cell within plant tissues during transpiration? a) Capillary action b) Osmosis c) Cohesion d) Cell elongation

52. How does transpiration in plants contribute to the movement of sugars within the phloem?
a) By creating a pressure gradient for sugar transport
b) By actively transporting sugars in the xylem
c) By increasing the solubility of sugars in the soil solution
d) By stimulating the release of sugars from the root hairs

53. Which of the following environmental factors can lead to a decrease in transpiration rate? a) High light intensity b) Low humidity c) High wind speed d) High temperature

54. How does the size and density of stomata affect transpiration rate in plants?
a) Smaller stomata decrease transpiration rate
e) Smaller stomata increase transpiration rate
f) Lower stomatal density decreases transpiration rate
d) Lower stomatal density increases transpiration rate

55. What is the term used to describe the movement of water from the soil into the roots of a plant during transpiration? a) Osmosis b) Photosynthesis c) Transpiration pull d) Capillary action

56. How does the structure of the leaf surface in plants affect transpiration rate?

a) A rough leaf surface increases transpiration rate
b) A smooth leaf surface increases transpiration rate
c) A waxy leaf surface decreases transpiration rate
d) A hairy leaf surface decreases transpiration rate

57. Which of the following environmental factors can lead to a decrease in transpiration rate? a) Low light intensity b) High humidity c) Low wind speed d) Low temperature

58. What is the term used to describe the process by which water moves from the xylem to the leaf surface during transpiration?
a) Osmosis
b) Capillary action
c) Transpiration pull
d) Evaporation

59. How does the presence of air pollutants, such as ozone, affect transpiration in plants?
a) It decreases transpiration rate
b) It increases transpiration rate
c) It has no effect on transpiration rate
d) It depends on the plant species

ANSWERS

1. c) The process of water loss through plant leaves

2. a) Leaves

3. d) Evaporation

4. a) Light intensity

5. c) Cooling of the plant

6. a) Guard cells

7. c) Transpiration

8. c) Water vapor

9. d) Bright sunlight

10. a) Nutrient absorption

11. c) Transpiration pull

12. c) Stomatal transpiration

13. d) Guard cell movement

14. b) It increases transpiration rate

15. c) Windy conditions

16. c) Water vapor

17. b) By releasing water vapor from the leaves

18. a) Roots

19. a) Transpiration

20. d) Soil pH

21. d) Evaporation

22. a) Wilting

23. d) Stomata

24. a) Cuticular transpiration

25. a) It increases transpiration rate

26. c) Roots

27. a) Cuticular transpiration

28. a) High humidity decreases transpiration rate

29. a) Leaves

30. d) Transpiration pull

31. a) High temperature increases transpiration rate

32. c) High soil moisture

33. c) Diffusion

34. c) Higher stomatal density increases transpiration rate

35. b) High humidity

36. b) It prevents the backflow of water in the xylem

37. a) Thick cuticle

38. a) Tracheids

39. a) It creates a concentration gradient for nutrient uptake

40. c) Humidity

41. a) Root pressure

42. a) It creates a pressure gradient for mineral absorption

43. d) Abscisic acid

44. b) A shallow root system increases transpiration rate

45. c) Osmosis

46. a) It decreases transpiration rate

47. d) High wind speed and low light intensity

48. a) Evaporation

49. a) By actively transporting nutrients in the xylem

50. a) Closing stomata during the day

51. c) Cohesion

52. a) By creating a pressure gradient for sugar transport

53. b) Low humidity

54. b) Smaller stomata increase transpiration rate

55. d) Capillary action

56. c) A waxy leaf surface decreases transpiration rate

57. b) High humidity

58. c) Transpiration pull

59. a) It decreases transpiration rate

PLANT HORMONES AND GROWTH REGULATORS

1. Which of the following is a plant hormone responsible for cell elongation and promoting stem growth? a) Auxin b) Cytokinin c) Gibberellin d) Ethylene

2. Which plant hormone is involved in the promotion of seed germination and breaking seed dormancy? a) Abscisic acid b) Auxin c) Cytokinin d) Gibberellin

3. Which plant hormone is responsible for controlling the opening and closing of stomata? a) Abscisic acid b) Auxin c) Cytokinin d) Gibberellin

4. Which plant hormone is involved in the promotion of lateral bud growth and inhibition of apical dominance? a) Auxin b) Cytokinin c) Gibberellin d) Ethylene

5. What is the function of gibberellins in plants?
a) Promotion of cell elongation and stem growth
b) Regulation of seed germination and dormancy
c) Control of leaf senescence and fruit ripening
d) Inhibition of lateral bud growth and promotion of apical dominance

6. Which plant hormone is responsible for promoting cell division and growth in the shoot and root meristems? a) Auxin b) Cytokinin c) Gibberellin d) Ethylene

7. Which plant hormone is responsible for promoting fruit ripening? a) Abscisic acid b) Auxin c) Cytokinin d) Ethylene

8. Which plant hormone is involved in the regulation of leaf abscission (leaf drop)? a) Abscisic acid b) Auxin c) Cytokinin d) Ethylene

9. What is the function of ethylene in plants?
a) Regulation of fruit ripening and senescence
b) Promotion of cell division and growth
c) Control of stomatal opening and closing
d) Inhibition of lateral bud growth and apical dominance

10. Which plant hormone is responsible for promoting root growth and branching? a) Auxin b) Cytokinin c) Gibberellin d) Ethylene

11. How do auxins affect plant growth?
a) They promote cell elongation and tropic responses.
b) They inhibit cell division and promote leaf senescence.
c) They regulate stomatal opening and closing.

d) They promote lateral bud growth and inhibit apical dominance.

12. Which plant hormone is responsible for regulating responses to stress, such as drought and salinity? a) Abscisic acid b) Auxin c) Cytokinin d) Gibberellin

13. Which plant hormone is involved in the control of phototropism, the growth of plants towards light? a) Abscisic acid b) Auxin c) Cytokinin d) Ethylene

14. What is the role of cytokinins in plants?
a) Promotion of cell division and shoot growth
b) Inhibition of seed germination and dormancy
c) Control of leaf senescence and fruit ripening
d) Inhibition of lateral bud growth and apical dominance

15. Which plant hormone is responsible for promoting the closing of stomata during water stress? a) Abscisic acid b) Auxin c) Cytokinin d) Gibberellin

16. How do gibberellins affect plant growth?
a) They promote cell elongation and stem growth.
b) They inhibit cell division and promote leaf senescence.
c) They regulate stomatal opening and closing.
d) They promote lateral bud growth and inhibit apical dominance.

17. Which plant hormone is involved in the regulation of plant defense responses, such as the formation of lignin and strengthening of cell walls?
a) Abscisic acid
b) Auxin
c) Cytokinin
d) Ethylene

18. What is the function of abscisic acid in plants?
a) Regulation of stomatal opening and closing
b) Promotion of cell division and shoot growth
c) Control of fruit ripening and senescence
d) Inhibition of lateral bud growth and promotion of apical dominance

19. Which plant hormone is responsible for promoting the elongation of cells on the shaded side of a plant stem, leading to bending towards light?
a) Abscisic acid
b) Auxin
c) Cytokinin
d) Gibberellin

20. How do ethylene affect plant growth?
a) It promotes fruit ripening and senescence.
b) It inhibits cell division and promotes leaf senescence.
c) It regulates stomatal opening and closing.
d) It promotes lateral bud growth and inhibits apical dominance.
21. Which plant hormone is responsible for promoting seed dormancy and inhibiting seed germination? a) Abscisic acid b) Auxin c) Cytokinin d) Ethylene

22. How do cytokinins affect plant growth?
a) They promote cell division and shoot growth.
b) They inhibit cell elongation and promote leaf senescence.
c) They regulate stomatal opening and closing.
d) They promote lateral bud growth and inhibit apical dominance.

23. Which plant hormone is responsible for the growth of adventitious roots? a) Abscisic acid b) Auxin c) Cytokinin d) Ethylene

24. What is the role of gibberellins in seed germination?
a) They break seed dormancy and promote germination.
b) They inhibit seed germination and promote dormancy.
c) They regulate stomatal opening and closing.
d) They promote lateral bud growth and inhibit apical dominance.

25. Which plant hormone is responsible for promoting the closing of stomata during water stress? a) Abscisic acid b) Auxin c) Cytokinin d) Ethylene

26. How do auxins affect plant growth?
a) They promote cell elongation and tropic responses.
b) They inhibit cell division and promote leaf senescence.
c) They regulate stomatal opening and closing.
d) They promote lateral bud growth and inhibit apical dominance.

27. What is the role of cytokinins in plants?
a) Promotion of cell division and shoot growth
b) Inhibition of seed germination and dormancy
c) Control of leaf senescence and fruit ripening
d) Inhibition of lateral bud growth and apical dominance

28. Which plant hormone is responsible for promoting the elongation of cells on the shaded side of a plant stem, leading to bending towards light? a) Abscisic acid b) Auxin c) Cytokinin d) Gibberellin

29. What is the function of gibberellins in plants?
a) Promotion of cell elongation and stem growth
b) Regulation of seed germination and dormancy
c) Control of leaf senescence and fruit ripening
d) Inhibition of lateral bud growth and promotion of apical dominance

30. Which plant hormone is involved in the control of phototropism, the growth of plants towards light? a) Abscisic acid b) Auxin c) Cytokinin d) Ethylene

31. What is the function of abscisic acid in plants?
a) Regulation of stomatal opening and closing
b) Promotion of cell division and shoot growth
c) Control of fruit ripening and senescence
d) Inhibition of lateral bud growth and promotion of apical dominance

32. How do ethylene affect plant growth?
a) It promotes fruit ripening and senescence.
b) It inhibits cell division and promotes leaf senescence.
c) It regulates stomatal opening and closing.
d) It promotes lateral bud growth and inhibits apical dominance.

33. Which plant hormone is responsible for promoting root growth and branching? a) Auxin b) Cytokinin c) Gibberellin d) Ethylene

34. How do gibberellins affect plant growth?
a) They promote cell elongation and stem growth.
b) They inhibit cell division and promote leaf senescence.
c) They regulate stomatal opening and closing.
d) They promote lateral bud growth and inhibit apical dominance.

35. What is the role of cytokinins in plants?
a) Promotion of cell division and shoot growth
b) Inhibition of seed germination and dormancy
c) Control of leaf senescence and fruit ripening
d) Inhibition of lateral bud growth and apical dominance

36. Which plant hormone is responsible for controlling the opening and closing of stomata? a) Abscisic acid b) Auxin c) Cytokinin d) Gibberellin

37. Which plant hormone is involved in the regulation of plant defense responses, such as the formation of lignin and strengthening of cell walls? a) Abscisic acid b) Auxin c) Cytokinin d) Ethylene

38. What is the function of auxins in plants?
a) Promotion of cell elongation and tropic responses
b) Inhibition of seed germination and dormancy
c) Control of leaf senescence and fruit ripening
d) Inhibition of lateral bud growth and apical dominance

39. Which plant hormone is responsible for promoting cell division and growth in the shoot and root meristems? a) Auxin b) Cytokinin c) Gibberellin d) Ethylene

40. How do auxins affect the growth of roots?
a) They promote root elongation and branching.
b) They inhibit root growth and promote shoot growth.
c) They regulate stomatal opening and closing.
d) They promote lateral bud growth and inhibit apical dominance.

41. What is the function of abscisic acid in plants?
a) Regulation of stomatal opening and closing
b) Promotion of cell division and shoot growth
c) Control of fruit ripening and senescence
d) Inhibition of lateral bud growth and promotion of apical dominance

42. Which plant hormone is responsible for promoting the closing of stomata during water stress? a) Abscisic acid b) Auxin c) Cytokinin d) Ethylene

43. How do cytokinins affect plant growth?
a) They promote cell division and shoot growth.
b) They inhibit cell elongation and promote leaf senescence.
c) They regulate stomatal opening and closing.
d) They promote lateral bud growth and inhibit apical dominance.

44. Which plant hormone is involved in the promotion of lateral bud growth and inhibition of apical dominance? a) Auxin b) Cytokinin c) Gibberellin d) Ethylene

45. How do ethylene affect plant growth?
a) It promotes fruit ripening and senescence.
b) It inhibits cell division and promotes leaf senescence.
c) It regulates stomatal opening and closing.
d) It promotes lateral bud growth and inhibits apical dominance.

46. What is the function of auxins in plants?
a) Promotion of cell elongation and tropic responses
b) Inhibition of seed germination and dormancy

c) Control of leaf senescence and fruit ripening

d) Inhibition of lateral bud growth and apical dominance

47. Which plant hormone is responsible for promoting seed dormancy and inhibiting seed germination? a) Abscisic acid b) Auxin c) Cytokinin d) Ethylene

48. How do cytokinins affect the growth of shoots?
a) They promote shoot elongation and branching.
b) They inhibit shoot growth and promote root growth.
c) They regulate stomatal opening and closing.
d) They promote lateral bud growth and inhibit apical dominance.

49. What is the role of gibberellins in plant reproduction?
a) They promote flower and fruit development.
b) They inhibit flowerdevelopment and promote leaf growth.
c) They regulate stomatal opening and closing.
d) They promote lateral bud growth and inhibit apical dominance.

50. Which plant hormone is responsible for promoting the closing of stomata during water stress? a) Abscisic acid b) Auxin c) Cytokinin d) Gibberellin

51. What is the function of jasmonic acid in plants?
a) Regulation of seed germination
b) Control of leaf senescence
c) Promotion of lateral bud growth
d) Induction of defense responses against herbivores

52. Which plant hormone is responsible for promoting cell division in the cambium and secondary growth? a) Auxin b) Cytokinin c) Ethylene d) Gibberellin

53. What is the role of brassinosteroids in plants? a) Promotion of seed germination b) Regulation of stomatal opening c) Control of leaf senescence d) Promotion of cell elongation and division

54. Which plant hormone is responsible for inhibiting seed germination and promoting dormancy? a) Abscisic acid b) Auxin c) Cytokinin d) Gibberellin

55. What is the function of strigolactones in plants? a) Promotion of lateral bud growth b) Control of leaf senescence c) Regulation of stomatal opening d) Induction of defense responses against pathogens

56. Which plant hormone is involved in the control of photoperiodic responses, such as flowering? a) Abscisic acid b) Auxin c) Cytokinin d) Gibberellin

57. What is the role of salicylic acid in plants? a) Regulation of seed germination b) Control of leaf senescence c) Induction of defense responses against pathogens d) Promotion of lateral bud growth

58. Which plant hormone is responsible for promoting the elongation of cells in response to gravity? a) Abscisic acid b) Auxin c) Cytokinin d) Ethylene

59. What is the function of abscisic acid in seed development?
a) Promotion of seed germination
b) Regulation of stomatal opening
c) Inhibition of seed germination and promotion of dormancy
d) Control of leaf senescence

60. Which plant hormone is involved in the control of leaf abscission (leaf drop)? a) Abscisic acid b) Auxin c) Cytokinin d) Gibberellin

61. What is the role of gibberellins in stem elongation? a) Promotion of cell division b) Inhibition of cell elongation c) Regulation of stomatal opening d) Promotion of cell elongation

62. Which plant hormone is responsible for promoting the closing of stomata during water stress? a) Abscisic acid b) Auxin c) Cytokinin d) Ethylene

63. What is the function of cytokinins in plant tissue culture? a) Promotion of root growth b) Inhibition of shoot elongation c) Control of leaf senescence d) Promotion of cell division

64. Which plant hormone is involved in the control of gravitropism, the growth response to gravity? a) Abscisic acid b) Auxin c) Cytokinin d) Gibberellin

65. What is the role of ethylene in fruit ripening? a) Inhibition of fruit ripening b) Control of stomatal opening c) Promotion of lateral bud growth d) Promotion of fruit ripening

66. Which plant hormone is responsible for promoting cell division in the root meristem? a) Auxin b) Cytokinin c) Ethylene d) Gibberellin

67. What is the function of abscisic acid in drought tolerance?
a) Promotion of root growth
b) Inhibition of stomatal opening

c) Control of leaf senescence
d) Promotion of lateral bud growth

68. Which plant hormone is involved in the control of apical dominance? a) Abscisic acid b) Auxin c) Cytokinin d) Gibberellin

69. What is the role of strigolactones in plant root development?
a) Promotion of lateral root growth
b) Inhibition of root elongation
c) Control of leaf senescence
d) Promotion of root hair formation

70. Which plant hormone is responsible for promoting cell division in the shoot meristem?
a) Auxin b) Cytokinin c) Ethylene d) Gibberellin

71. What is the function of abscisic acid in seed dormancy? a) Promotion of seed germination b) Inhibition of stomatal opening c) Control of leaf senescence d) Promotion of seed dormancy

72. Which plant hormone is involved in the control of phototropism, the growth response to light? a) Abscisic acid b) Auxin c) Cytokinin d) Gibberellin

73. What is the role of jasmonic acid in plant defense responses?
a) Promotion of lateral bud growth
b) Control of leaf senescence
c) Induction of defense responses against pathogens and herbivores
d) Regulation of stomatal opening

74. Which plant hormone is responsible for promoting root elongation and gravitropic responses? a) Abscisic acid b) Auxin c) Cytokinin d) Ethylene

75. What is the function of brassinosteroids in plant development? a) Promotion of seed germination b) Inhibition of leaf growth c) Control of leaf senescence d) Promotion of cell elongation and division

76. Which plant hormone is involved in the control of leaf senescence (aging)? a) Abscisic acid b) Auxin c) Cytokinin d) Gibberellin

ANSWERS

1. a) Auxin

2. d) Gibberellin

3. a) Abscisic acid

4. b) Cytokinin

5. a) Promotion of cell elongation and stem growth

6. b) Cytokinin

7. d) Ethylene

8. a) Abscisic acid

9. a) Regulation of fruit ripening and senescence

10. a) Auxin

11. a) They promote cell elongation and tropic responses.

12. a) Abscisic acid

13. b) Auxin

14. a) Promotion of cell division and shoot growth

15. a) Abscisic acid

16. a) They promote cell elongation and stem growth.

17. d) Ethylene

18. a) Regulation of stomatal opening and closing

19. b) Auxin

20. a) It promotes fruit ripening and senescence.

21. a) Abscisic acid

22. a) They promote cell division and shoot growth.

23. b) Auxin

24. a) They break seed dormancy and promote germination.

25. a) Abscisic acid

26. a) They promote cell elongation and tropic responses.

27. a) Promotion of cell division and shoot growth

28. b) Auxin

29. a) Promotion of cell elongation and stem growth

30. b) Auxin

31. a) Regulation of stomatal opening and closing

32. a) It promotes fruit ripening and senescence.

33. a) Auxin

34. a) They promote cell elongation and stem growth.

35. a) Promotion of cell division and shoot growth

36. b) Auxin

37. d) Ethylene

38. a) Promotion of cell elongation and tropic responses

39. a) Auxin

40. a) They promote root elongation and branching.

41. a) Regulation of stomatal opening and closing

42. a) Abscisic acid

43. a) They promote cell division and shoot growth.

44. a) Auxin

45. a) It promotes fruit ripening and senescence.

46. a) Promotion of cell elongation and tropic responses

47. a) Abscisic acid

48. a) They promote shoot elongation and branching.

49. a) They promote flower and fruit development.

50. a) Abscisic acid

51. d) Induction of defense responses against herbivores

52. b) Cytokinin

53. d) Promotion of cell elongation and division

54. a) Abscisic acid

55. a) Promotion of lateral bud growth

56. d) Gibberellin

57. c) Induction of defense responses against pathogens

58. b) Auxin

59. c) Inhibition of seed germination and promotion of dormancy

60. a) Abscisic acid

61. d) Promotion of cell elongation

62. a) Abscisic acid

63. d) Promotion of cell division

64. b) Auxin

65. d) Promotion of fruit ripening

66. b) Cytokinin

67. b) Inhibition of stomatal opening

68. b) Auxin

69. a) Promotion of lateral root growth

70. b) Cytokinin

71. d) Promotion of seed dormancy

72. b) Auxin

73. c) Induction of defense responses against pathogens and herbivores

74. b) Auxin

75. d) Promotion of cell elongation and division

76. c) Cytokinin

PLANT NUTRITION AND METABOLISM

1. Which of the following is not considered a macronutrient for plants? a) Nitrogen b) Iron c) Potassium d) Phosphorus

2. Which process in plants involves the conversion of light energy into chemical energy? a) Photosynthesis b) Transpiration c) Respiration d) Germination

3. Which of the following nutrients is involved in the formation of chlorophyll? a) Zinc b) Magnesium c) Manganese d) Copper

4. Which part of the plant is primarily responsible for the absorption of water and minerals? a) Roots b) Leaves c) Stem d) Flowers

5. In which form do plants primarily absorb nitrogen?
a) Nitrate (NO_3^-)
b) Ammonium (NH_4^+)
c) Nitrogen gas (N_2)
d) Nitrite (NO_2^-)

6. Which of the following is a characteristic of C4 plants?
a) They use the C3 photosynthetic pathway
b) They fix carbon dioxide through the Calvin cycle
c) They have bundle sheath cells with high carbon dioxide concentration
d) They are adapted to hot and dry conditions

7. What is the primary function of xylem tissue in plants? a) Translocation of sugars b) Transport of water and minerals c) Storage of starch d) Support and protection

8. Which of the following is a micronutrient required by plants in small amounts? a) Nitrogen b) Phosphorus c) Potassium d) Zinc

9. Which of the following is responsible for the breakdown of glucose in plant cells? a) Mitochondria b) Chloroplasts c) Ribosomes d) Vacuoles

10. Which of the following is a symptom of iron deficiency in plants? a) Leaf wilting b) Leaf yellowing (chlorosis) c) Stunted root growth d) Leaf curling

11. Which of the following is the primary source of energy for plant metabolism? a) Glucose b) Sucrose c) Starch d) Fructose

12. Which process involves the movement of sugars from source to sink in plants? a) Transpiration b) Translocation c) Respiration d) Photosynthesis

13. Which of the following is not a factor that affects the rate of photosynthesis? a) Light intensity b) Temperature c) Oxygen concentration d) Carbon dioxide concentration

14. Which molecule is the final electron acceptor in the electron transport chain during photosynthesis? a) NADPH b) ATP c) Oxygen d) Carbon dioxide

15. Which of the following is the primary pigment involved in photosynthesis? a) Chlorophyll a b) Chlorophyll b c) Carotenoids d) Xanthophylls

16. Which of the following is not a function of auxins in plants?
a) Promoting cell elongation
b) Inhibiting lateral bud growth
c) Stimulating root development
d) Enhancing leaf senescence

17. Which of the following is an example of an essential micronutrient for plant growth? a) Sodium b) Calcium c) Copper d) Sulphur

18. In which part of the plant does the process of nitrogen fixation occur? a) Roots b) Stems c) Leaves d) Nodules

19. Which of the following is not a component of the photosynthetic electron transport chain? a) Photosystem I b) Photosystem II c) Cytochrome complex d) Glycolysis

20. Which of the following is a parasitic plant that lacks chlorophyll and obtains nutrients from host plants? a) Sunflower b) Venus flytrap c) Dodder d) Dandelion

21. Which of the following is the primary nitrogenous waste product excreted by plants? a) Urea b) Ammonia c) Uric acid d) Creatinine

22. Which of the following processes requires oxygen and produces carbon dioxide in plant cells? a) Photosynthesis b) Transpiration c) Respiration d) Fermentation

23. What is the primary role of gibberellins in plant growth and development?
a) Promoting seed germination
b) Inhibiting root growth
c) Enhancing leaf senescence
d) Stimulating flower abscission

24. Which of the following is not an essential element required for plant growth? a) Carbon b) Hydrogen c) Oxygen d) Chlorine

25. Which organelle is responsible for the synthesis of ATP during cellular respiration in plants? a) Mitochondria b) Chloroplasts c) Golgi apparatus d) Endoplasmic reticulum

26. Which process involves the breakdown of glucose to release energy in the absence of oxygen? a) Photosynthesis b) Fermentation c) Transpiration d) Translocation

27. Which of the following is not a mobile nutrient in plants? a) Nitrogen b) Potassium c) Magnesium d) Iron

28. What is the primary role of cytokinins in plant growth and development?
a) Promoting cell elongation
b) Inhibiting lateral bud growth
c) Stimulating root development
d) Delaying leaf senescence

29. Which of the following is not a form of passive transport in plant cells? a) Diffusion b) Osmosis c) Active transport d) Facilitated diffusion

30. Which of the following is an example of a plant hormone involved in gravitropism? a) Auxin b) Gibberellin c) Abscisic acid d) Ethylene

31. Which of the following elements is involved in the synthesis of amino acids, proteins, and nucleic acids in plants? a) Sulfur b) Calcium c) Boron d) Cobalt

32. Which of the following pigments is responsible for the red coloration in autumn leaves? a) Chlorophyll a b) Chlorophyll b c) Anthocyanins d) Carotenoids

33. In which part of the chloroplast do the light-dependent reactions of photosynthesis occur? a) Stroma b) Thylakoid membrane c) Outer membrane d) Granum

34. Which of the following is not a metabolic pathway involved in the breakdown of glucose in plant cells?
a) Glycolysis
b) Krebs cycle (citric acid cycle)
c) Calvin cycle (carbon fixation)
d) Electron transport chain

35. Which of the following elements is required for the synthesis of enzymes involved in nitrogen fixation? a) Molybdenum b) Silicon c) Cobalt d) Vanadium

36. Which of the following hormones promotes stomatal closure in response to water stress? a) Abscisic acid (ABA) b) Gibberellin c) Cytokinin d) Ethylene

37. Which of the following is an example of a C3 plant? a) Corn (maize) b) Sugarcane c) Pineapple d) Rice

38. Which of the following is the primary site of water absorption in a root hair? a) Casparian strip b) Epidermis c) Endodermis d) Cortex

39. Which of the following is a phenomenon where plants bend towards a light source? a) Geotropism b) Phototropism c) Thigmotropism d) Hydrotropism

40. Which of the following is an example of an essential micronutrient for plant growth? a) Boron b) Sodium c) Cobalt d) Silicon

41. Which of the following is a characteristic of CAM plants?
a) They have specialized bundle sheath cells
b) They open stomata during the day
c) They fix carbon dioxide through the C4 pathway
d) They are adapted to cool and moist environments

42. Which of the following is the primary source of carbon for plants during photosynthesis?
a) Carbon dioxide (CO_2)
b) Oxygen (O_2)
c) Water (H_2O)
d) Glucose ($C_6H_{12}O_6$)

43. Which of the following hormones is responsible for apical dominance in plants? a) Auxin b) Gibberellin c) Abscisic acid (ABA) d) Ethylene

44. Which of the following is the process of converting nitrates into nitrogen gas by bacteria in the soil? a) Nitrogen fixation b) Nitrification c) Denitrification d) Ammonification

45. Which of the following pigments is responsible for the yellow coloration in flowers and fruits? a) Chlorophyll a b) Chlorophyll b c) Anthocyanins d) Xanthophylls

46. Which of the following is the primary function of phloem tissue in plants?
a) Transport of water and minerals
b) Storage of starch
c) Support and protection

d) Translocation of sugars

47. Which of the following is a process that involves the conversion of nitrogen gas into nitrates by certain bacteria?
a) Nitrogen fixation
b) Nitrification
c) Denitrification
d) Ammonification

48. Which of the following is not a function of cytokinins in plants?
a) Promoting cell division
b) Inhibiting leaf senescence
c) Stimulating lateral bud growth
d) Enhancing root development

49. Which of the following is the primary location of the enzyme rubisco in a chloroplast?
a) Stroma b) Thylakoid membrane c) Outer membrane d) Granum

50. Which of the following is the primary mechanism by which plants take up water from the soil? a) Active transport b) Facilitated diffusion c) Osmosis d) Bulk flow

51. Which element is essential for the synthesis of chlorophyll? a) Nitrogen b) Phosphorus c) Iron d) Magnesium

52. Which of the following is not a macronutrient for plants? a) Nitrogen b) Potassium c) Zinc d) Calcium

53. What is the process by which plants convert light energy into chemical energy? a) Photosynthesis b) Respiration c) Transpiration d) Absorption

54. Which of the following nutrients is responsible for promoting root development and fruit formation? a) Phosphorus b) Iron c) Manganese d) Copper

55. What is the primary function of micronutrients in plants?
a) They are required in large quantities for growth.
b) They play a role in photosynthesis.
c) They act as catalysts for various metabolic reactions.
d) They provide structural support to plant cells.

56. What is the primary function of nitrogen in plants?
a) It promotes root development.
b) It plays a role in photosynthesis.

c) It is involved in DNA replication.
d) It is required for the synthesis of amino acids and proteins.

57. Which of the following elements is responsible for activating enzymes involved in energy transfer reactions in plants? a) Potassium b) Magnesium c) Boron d) Zinc

58. What is the primary function of phosphorus in plants?
a) It promotes leaf growth.
b) It is involved in the synthesis of DNA and RNA.
c) It is required for the synthesis of chlorophyll.
d) It plays a role in water and nutrient uptake.

59. Which of the following is not a mode of nutrient uptake by plants?
a) Absorption through roots
b) Assimilation through leaves
c) Direct synthesis through photosynthesis
d) Symbiotic association with mycorrhizal fungi

60. Which of the following is not a source of organic carbon for plants?
a) CO_2 in the atmosphere
b) Organic matter in the soil
c) Photosynthesis
d) Decomposition of organic material

61. Which process involves the breakdown of complex organic molecules into simpler ones to release energy? a) Photosynthesis b) Respiration c) Transpiration d) Assimilation

62. What is the primary function of potassium in plants?
a) It promotes flowering and fruiting.
b) It is involved in the synthesis of chlorophyll.
c) It plays a role in water and nutrient uptake.
d) It is required for the synthesis of amino acids and proteins.

63. Which of the following nutrients is responsible for activating enzymes involved in photosynthesis? a) Manganese b) Nitrogen c) Calcium d) Sulfur

64. What is the process by which plants absorb water and dissolved nutrients from the soil? a) Transpiration b) Photosynthesis c) Respiration d) Root uptake

65. Which of the following is not a mobile nutrient in plants? a) Nitrogen b) Phosphorus c) Potassium d) Magnesium

66. What is the primary function of iron in plants?
a) It promotes root growth.
b) It is involved in the synthesis of DNA and RNA.
c) It plays a role in water and nutrient uptake.
d) It is required for the synthesis of chlorophyll.

67. Which of the following elements is a component of nucleic acids and ATP in plants?
   a) Phosphorus b) Iron c) Manganese d) Copper

68. What is the primary function of calcium in plants?
a) It promotes leaf growth.
b) It is involved in the synthesis of DNA and RNA.
c) It plays a role in water and nutrient uptake.
d) It provides structural support to plant cells.

69. Which of the following is an essential macronutrient for plants that is often supplied through fertilizers? a) Zinc b) Boron c) Sulfur d) Calcium

70. What is the primary function of magnesium in plants?
a) It promotes root development.
b) It is involved in the synthesis of DNA and RNA.
c) It plays a role in water and nutrient uptake.
d) It is required for the synthesis of chlorophyll.

ANSWERS

1. b) Iron
2. a) Photosynthesis
3. b) Magnesium
4. a) Roots
5. a) Nitrate (NO3-)
6. c) They have bundle sheath cells with high carbon dioxide concentration
7. b) Transport of water and minerals
8. d) Zinc
9. a) Mitochondria
10. b) Leaf yellowing (chlorosis)
11. a) Glucose
12. b) Translocation
13. c) Oxygen concentration
14. c) Oxygen
15. a) Chlorophyll a
16. d) Enhancing leaf senescence
17. c) Copper
18. d) Nodules
19. d) Glycolysis
20. c) Dodder
21. b) Ammonia
22. c) Respiration
23. a) Promoting seed germination
24. d) Chlorine
25. a) Mitochondria
26. b) Fermentation
27. d) Iron
28. d) Delaying leaf senescence
29. c) Active transport

30. a) Auxin

31. a) Sulfur

32. c) Anthocyanins

33. b) Thylakoid membrane

34. c) Calvin cycle (carbon fixation)

35. a) Molybdenum

36. a) Abscisic acid (ABA)

37. d) Rice

38. b) Epidermis

39. b) Phototropism

40. a) Boron

41. b) They open stomata during the day

42. a) Carbon dioxide (CO2)

43. a) Auxin

44. c) Denitrification

45. d) Xanthophylls

46. d) Translocation of sugars

47. a) Nitrogen fixation

48. c) Stimulating lateral bud growth

49. a) Stroma

50. c) Osmosis

51. d) Magnesium

52. c) Zinc

53. a) Photosynthesis

54. a) Phosphorus

55. c) They act as catalysts for various metabolic reactions.

56. d) It is required for the synthesis of amino acids and proteins.

57. b) Magnesium

58. b) It is involved in the synthesis of DNA and RNA.

59. c) Direct synthesis through photosynthesis

60. c) Photosynthesis

61. b) Respiration

62. a) It promotes flowering and fruiting.

63. a) Manganese

64. d) Root uptake

65. d) Magnesium

66. d) It is required for the synthesis of chlorophyll.

67. a) Phosphorus

68. d) It provides structural support to plant cells.

69. c) Sulfur

70. d) It is required for the synthesis of chlorophyll.

CHAPTER 3: PLANT REPRODUCTION AND DEVELOPMENT

SEXUAL AND ASEXUAL REPRODUCTION

1. Which of the following is a characteristic of sexual reproduction in plants?
a. Involves the fusion of male and female gametes
b. Only requires one parent plant
c. Produces genetically identical offspring
d. Occurs through vegetative propagation

2. Which structure is responsible for the production of pollen in plants? a. Anther b. Ovary c. Stigma d. Style

3. In angiosperms, the male gametophyte is commonly referred to as: a. Pollen b. Pistil c. Stamen d. Sepal

4. Which of the following is an example of asexual reproduction in plants? a. Seed production b. Spore formation c. Vegetative propagation d. Cross-pollination

5. A plant that reproduces asexually by forming new shoots from its roots is utilizing: a. Rhizomes b. Stolons c. Tubers d. Bulbs

6. Which type of asexual reproduction occurs when a new plantlet develops from the parent plant and then detaches? a. Fragmentation b. Budding c. Runners d. Offsets

7. The transfer of pollen from the anther to the stigma of the same flower is called: a. Self-pollination b. Cross-pollination c. Wind pollination d. Insect pollination

8. Which of the following is a characteristic of asexual reproduction in plants?
a. Results in genetic variation
b. Requires pollinators
c. Involves the fusion of gametes
d. Produces genetically identical offspring

9. What is the purpose of the ovary in plants? a. Produces pollen grains b. Attracts pollinators c. Protects the ovules d. Releases sperm cells

10. Which type of sexual reproduction involves the fusion of male and female gametes within a flower? a. Fertilization b. Spore formation c. Fragmentation d. Budding

11. Asexual reproduction in plants does not involve: a. Pollination b. Seeds c. Flowers d. Spores

12. Which of the following is a method of asexual reproduction in fungi? a. Budding b. Cross-pollination c. Self-pollination d. Wind dispersal

13. In plants, the female reproductive structure is called: a. Pistil b. Stamen c. Sepal d. Anther

14. The process of producing new individuals from the vegetative parts of a plant is known as: a. Vegetative propagation b. Fertilization c. Cross-pollination d. Spore formation

15. Which of the following is a type of asexual reproduction in which a new plant grows from a cut stem or leaf? a. Cutting b. Grafting c. Layering d. Bulbil formation

16. The transfer of pollen from the anther of one flower to the stigma of another flower is called: a. Self-pollination b. Cross-pollination c. Wind pollination d. Insect pollination

17. What is the male gamete in plants called? a. Pollen b. Ovule c. Embryo d. Seed

18. Asexual reproduction is common in plants that live in environments where:
a. Pollinators are abundant
b. Genetic variation is important
c. Water is scarce
d. Seeds can be easily dispersed

19. Which of the following is a method of asexual reproduction in ferns? a. Rhizomes b. Stolons c. Tubers d. Bulbs

20. What is the purpose of the stigma in plants? a. Produces pollen grains b. Attracts pollinators c. Protects the ovules d. Receives pollen grains

21. Asexual reproduction in plants can occur through: a. Vegetative propagation b. Seed production c. Cross-pollination d. Self-pollination

22. Which of the following is an example of asexual reproduction in plants that involves the development of new plants from modified stems that run along the ground? a. Runners b. Rhizomes c. Stolons d. Bulbs

23. What is the purpose of the anther in plants? a. Produces pollen grains b. Attracts pollinators c. Protects the ovules d. Receives pollen grains

24. Sexual reproduction in plants contributes to: a. Genetic diversity b. Genetic stability c. Decreased fertility d. Rapid population growth

25. Which of the following is a type of asexual reproduction in which a new plant is produced from a specialized bud? a. Bulbil formation b. Grafting c. Layering d. Cutting

26. In plants, what is the structure that contains the ovules? a. Ovary b. Anther c. Stigma d. Style

27. Which of the following is a method of asexual reproduction in algae? a. Fragmentation b. Budding c. Spore formation d. Self-pollination

28. What is the purpose of the style in plants? a. Produces pollen grains b. Attracts pollinators c. Protects the ovules d. Receives pollen grains

29. Asexual reproduction is advantageous to plants in environments where: a. Pollinators are scarce b. Genetic variation is important c. Water is abundant d. Seeds can be easily dispersed

30. What is the female gamete in plants called? a. Ovule b. Pollen c. Embryo d. Seed

31. Which type of sexual reproduction involves the fusion of male and female gametes within the same organism? a. Self-fertilization b. Cross-fertilization c. External fertilization d. Internal fertilization

32. Asexual reproduction in plants is particularly advantageous when: a. Environmental conditions are stable b. Genetic diversity is required c. Seeds can be easily dispersed d. Pollinators are abundant

33. Which of the following is a method of asexual reproduction in bryophytes (mosses)? a. Rhizomes b. Stolons c. Gemmae cups d. Bulbs

34. What is the purpose of the sepals in plants? a. Produces pollen grains b. Attracts pollinators c. Protects the ovules d. Encloses and protects the flower bud

35. The process of fusion of male and female gametes to form a zygote is called: a. Fertilization b. Pollination c. Germination d. Maturation

36. Asexual reproduction in plants allows for rapid: a. Genetic variation b. Population growth c. Development of flowers d. Seed dispersal

37. Which of the following is a method of asexual reproduction in fungi where a small outgrowth detaches and develops into a new organism?
a. Budding
b. Fragmentation
c. Spore formation
d. Self-pollination

38. What is the purpose of the stamen in plants? a. Produces pollen grains b. Attracts pollinators c. Protects the ovules d. Receives pollen grains

39. Which type of sexual reproduction involves the fusion of male and female gametes outside the organism's body, typically in water?
a. External fertilization
b. Internal fertilization
c. Self-fertilization
d. Cross-fertilization

40. Which of the following is an example of asexual reproduction in plants that involves the development of new plants from underground stems? a. Rhizomes b. Runners c. Stolons d. Tubers

41. What is the purpose of the ovule in plants? a. Produces pollen grains b. Attracts pollinators c. Protects the embryo d. Receives pollen grains

42. Sexual reproduction in plants ensures: a. Genetic stability b. Decreased fertility c. Rapid population growth d. Genetic diversity

43. Which of the following is a type of asexual reproduction in which a new plant is formed from a lateral branch that touches the ground and takes root? a. Layering b. Grafting c. Cutting d. Bulbil formation

44. In plants, what is the structure that connects the stigma to the ovary? a. Style b. Anther c. Sepal d. Ovary

45. Asexual reproduction in plants allows for the production of: a. Genetic variation b. Vegetative parts c. Seeds d. Flowers

46. Which of the following is a method of asexual reproduction in bryophytes (mosses)? a. Rhizoids b. Stolons c. Gemmae cups d. Bulbs

47. What is the purpose of the petals in plants? a. Produces pollen grains b. Attracts pollinators c. Protects the ovules d. Encloses and protects the flower bud

48. The fusion of male and female gametes in plants results in the formation of a: a. Seed b. Embryo c. Pollen grain d. Sepal

49. Asexual reproduction in plants can occur through: a. Fragmentation b. Cross-pollination c. Self-pollination d. Wind dispersal

50. In gymnosperms, what is the structure that contains the ovules? a) Ovary b) Anther c) Stigma d) Cone

51. Which of the following is a method of asexual reproduction in plants where a new plant is formed from a portion of the parent plant breaking off? a) Fragmentation b) Bulbil formation c) Grafting d) Cutting

52. What is the purpose of the style in plants? a) Produces pollen grains b) Attracts pollinators c) Protects the ovules d) Receives pollen grains

53. Which type of sexual reproduction involves the fusion of male and female gametes within the same flower? a) Self-fertilization b) Cross-fertilization c) External fertilization d) Internal fertilization

54. What is the term used to describe the fusion of male and female gametes outside the organism's body, typically in water? a) External fertilization b) Internal fertilization c) Self-fertilization d) Cross-fertilization

55. Which of the following is a method of asexual reproduction in plants that involves the production of specialized structures that detach and develop into new plants? a) Adventitious roots b) Tubers c) Stolons d) Bulbils

56. What is the purpose of the petal in plants? a) Produces pollen grains b) Attracts pollinators c) Protects the ovules d) Encloses and protects the flower bud

57. Which of the following is a type of asexual reproduction in plants that involves the production of specialized structures called gemmae? a) Rhizomes b) Stolons c) Gemmae cups d) Bulbs

58. In ferns, what is the structure that contains clusters of sporangia called? a) Sori b) Strobili c) Cones d) Catkins

59. What is the purpose of the seed coat in plants? a) Protection of the embryo b) Attraction of pollinators c) Storage of nutrients d) Seed dispersal

60. Which type of sexual reproduction involves the fusion of male and female gametes outside the organism's body, facilitated by wind or water? a) Anemophily b) Hydrophily c) Zoophily d) Cleistogamy

61. Which of the following is a method of asexual reproduction in plants that involves the formation of new plants from underground stems? a) Rhizomes b) Runners c) Offset d) Bulbils

62. What is the purpose of the carpel in plants? a) Produces pollen grains b) Attracts pollinators c) Protects the ovules d) Receives pollen grains

63. Which type of sexual reproduction involves the fusion of gametes within the body of the organism? a) Internal fertilization b) External fertilization c) Self-fertilization d) Cross-fertilization

64. What is the structure in gymnosperms that contains the male gametophyte? a) Pollen b) Ovule c) Embryo d) Seed

65. Which of the following is a type of asexual reproduction in plants that involves the formation of new plants from specialized underground stems? a) Rhizomes b) Tubers c) Corms d) Stolons

66. In bryophytes (mosses), what is the structure that produces egg cells called? a) Antheridia b) Archegonia c) Sporophytes d) Gametophytes

67. What is the purpose of the nectar in flowers? a) Produces pollen grains b) Attracts pollinators c) Protects the ovules d) Encloses and protects the flower bud

68. Which of the following is a method of asexual reproduction in plants that involves the production of new plants from modified stems that grow horizontally above the ground? a) Rhizomes b) Tubers c) Stolons d) Corms

69. In angiosperms, what is the female gametophyte called? a) Embryo b) Ovule c) Pollen d) Seed

ANSWERS

1. a) Involves the fusion of male and female gametes

2. a) Anther

3. a) Pollen

4. c) Vegetative propagation

5. a) Rhizomes

6. c) Runners

7. a) Self-pollination

8. d) Produces genetically identical offspring

9. c) Protects the ovules

10. a) Fertilization

11. c) Flowers

12. a) Budding

13. a) Pistil

14. a) Vegetative propagation

15. a) Cutting

16. b) Cross-pollination

17. a) Pollen

18. c) Water is scarce

19. a) Rhizomes

20. d) Receives pollen grains

21. a) Vegetative propagation

22. a) Runners

23. a) Produces pollen grains

24. a) Genetic diversity

25. a) Bulbil formation

26. a) Ovary

27. a) Fragmentation

28. d) Receives pollen grains

29. a) Pollinators are scarce

30. a) Ovule

31. a) Self-fertilization

32. a) Environmental conditions are stable

33. c) Gemmae cups

34. d) Encloses and protects the flower bud

35. a) Fertilization

36. b) Population growth

37. a) Budding

38. a) Produces pollen grains

39. a) External fertilization

40. a) Rhizomes

41. c) Protects the embryo

42. d) Genetic diversity

43. a) Layering

44. a) Style

45. b) Vegetative parts

46. c) Gemmae cups

47. b) Attracts pollinators

48. b) Embryo

49. a) Fragmentation

50. a) Ovary

51. a) Fragmentation

52. d) Receives pollen grains

53. a) Self-fertilization

54. a) External fertilization

55. d) Bulbils

56. b) Attracts pollinators

57. c) Gemmae cups

58. a) Sori

59. a) Protection of the embryo

60. b) Hydrophily
61. a) Rhizomes
62. c) Protects the ovules
63. a) Internal fertilization
64. a) Pollen
65. a) Rhizomes
66. b) Archegonia
67. b) Attracts pollinators
68. c) Stolons
69. b) Ovule

POLLINATION AND FERTILIZATION

1. Pollination is the process of:
a) Transferring pollen from the stigma to the ovule
b) Transferring pollen from the anther to the stigma
d) Transferring sperm cells from the pollen to the ovule
d) Transferring nutrients from the style to the ovary

2. Which of the following is a common agent of pollination? a) Wind b) Water c) Gravity d) Fungi

3. The transfer of pollen from the anther to the stigma within the same flower is called: a) Self-pollination b) Cross-pollination c) Wind pollination d) Insect pollination

4. Which of the following is NOT a method of pollination? a) Insect pollination b) Wind pollination c) Self-pollination d) Fruit pollination

5. In self-pollination, the pollen is transferred:
a) From one flower to another flower of the same plant
b) From one plant to another plant of the same species
c) From the anther to the stigma of the same flower
d) From the stigma to the ovary of the same flower

6. Fertilization in plants involves the fusion of: a) Pollen and ovule b) Pollen and stigma c) Ovule and ovary d) Stigma and style

7. Where does fertilization occur in flowering plants? a) Anther b) Stigma c) Ovary d) Petal

8. After fertilization, the ovule develops into a(n): a) Embryo b) Stamen c) Sepal d) Pollen grain

9. The male gamete in plants is contained within the: a) Pollen grain b) Ovule c) Embryo d) Stigma

10. Double fertilization in angiosperms involves the fusion of one sperm cell with the egg cell to form the zygote, and the other sperm cell fuses with the _____ to form the endosperm. a) Ovary b) Style c) Polar nuclei d) Sepal

11. Which of the following is an example of a plant adaptation that promotes cross-pollination?
a) Production of large amounts of nectar

b) Development of self-incompatibility mechanisms
c) Release of pollen in the absence of pollinators
d) Production of numerous self-pollinated flowers

12. In plants, the transfer of pollen from the anther to the stigma is known as: a) Double fertilization b) Pollen germination c) Pollination d) Fertilization

13. What is the purpose of the pollen tube during fertilization in plants?
a) Provides structural support to the anther
b) Ensures the development of multiple embryos
c) Transports sperm cells to the ovule
d) Secures the attachment of the ovary to the receptacle

14. Which of the following is an example of a plant with a specialized pollination mechanism? a) Sunflower b) Dandelion c) Lily of the valley d) Tomato

15. Some plants have evolved mechanisms to prevent self-fertilization. This is known as: a) Cleistogamy b) Dichogamy c) Geitonogamy d) Self-incompatibility

16. In gymnosperms, the male gametes are typically contained within: a) Pollen grains b) Ovules c) Ovaries d) Anthers

17. Which of the following is NOT a common vector for pollination? a) Bees b) Butterflies c) Hummingbirds d) Earthworms

18. What is the primary role of the stigma in the process of pollination? a) Production of pollen grains b) Attraction of pollinators c) Protection of the ovules d) Reception of pollen grains

19. What is the term for plants that require a specific pollinator species to successfully reproduce? a) Generalist plants b) Specialist plants c) Self-pollinating plants d) Wind-pollinated plants

20. In gymnosperms, the transfer of pollen to the ovule occurs through the: a) Stigma b) Ovary c) Micropyle d) Embryo sac

21. What is the process by which a pollen grain lands on the stigma and begins to grow a pollen tube called? a) Germination b) Maturation c) Fertilization d) Dispersion

22. In plants, the fusion of one sperm cell with the egg cell results in the formation of the: a) Zygote b) Endosperm c) Seed coat d) Embryo sac

23. What is the term for the transfer of pollen from the anther to the stigma of a different flower on the same plant? a) Self-pollination b) Autogamy c) Chasmogamy d) Homogamy

24. What is the advantage of self-pollination in plants? a) Increased genetic diversity b) Greater potential for outbreeding c) Conservation of energy and resources d) Enhanced reproductive success

25. In wind-pollinated plants, pollen grains are often: a) Large and sticky b) Small and smooth c) Brightly colored d) Produced in abundance

26. What is the name of the structure that connects the stigma to the ovary in a flower? a) Style b) Stamen c) Sepal d) Filament

27. In plants, the stigma is covered by a waxy substance called: a) Cuticle b) Calyx c) Style d) Exine

28. What is the term for the transfer of pollen from the anther of one flower to the stigma of a flower on a different plant? a) Cross-pollination b) Xenogamy c) Heterogamy d) Hybridization

29. Which of the following is a type of pollination that occurs at night? a) Diurnal pollination b) Nocturnal pollination c) Entomophilous pollination d) Anemophilous pollination

30. In plants, what is the process by which the male gametophyte reaches the ovule called? a) Pollination b) Fertilization c) Germination d) Imbibition

31. In certain orchids, the structure that resembles a bee or other insect and attracts pollinators is called a: a) Nectary b) Stamen c) Petal d) Labellum

32. Some plants produce extra-floral nectaries, which serve as a food reward for: a) Pollinators b) Seed dispersers c) Herbivores d) Parasitic plants

33. What is the term for the process of transferring pollen from the anther to the stigma of a different flower on a different plant? a) Allogamy b) Cross-pollination c) Outcrossing d) Xenogamy

34. In plants, what is the process by which the pollen tube grows through the style toward the ovary called? a) Pollination b) Germination c) Fertilization d) Tube elongation

35. Some plants have evolved mechanisms to ensure that pollen is transferred between different flowers of the same individual. This is known as: a) Geitonogamy b) Dichogamy c) Homogamy d) Cleistogamy

36. What is the term for the fusion of a sperm cell with the egg cell during fertilization in plants? a) Double fertilization b) Syngamy c) Gametogenesis d) Embryogenesis

37. In flowering plants, what is the purpose of the endosperm? a) Protection of the embryo b) Attraction of pollinators c) Storage of nutrients d) Seed dispersal

38. Which of the following is a mechanism that promotes outcrossing in plants? a) Self-pollination b) Apomixis c) Homozygosity d) Self-incompatibility

39. In gymnosperms, the male gametophyte is commonly referred to as: a) Pollen b) Ovule c) Embryo d) Seed

40. What is the purpose of the anther in plants? a) Produces pollen grains b) Attracts pollinators c) Protects the ovules d) Receives pollen grains

ANSWERS

1. b) Transferring pollen from the anther to the stigma
2. a) Wind
3. a) Self-pollination
4. d) Fruit pollination
5. c) From the anther to the stigma of the same flower
6. a) Pollen and ovule
7. c) Ovary
8. a) Embryo
9. a) Pollen grain
10. c) Polar nuclei
11. a) Production of large amounts of nectar
12. a) Double fertilization
13. c) Transports sperm cells to the ovule
14. c) Lily of the valley
15. d) Self-incompatibility
16. a) Pollen grains
17. d) Earthworms
18. d) Reception of pollen grains
19. b) Specialist plants
20. c) Micropyle
21. a) Germination
22. a) Zygote
23. a) Self-pollination
24. c) Conservation of energy and resources
25. b) Small and smooth
26. a) Style
27. a) Cuticle
28. b) Cross-pollination
29. b) Nocturnal pollination

30. a) Pollination
31. d) Labellum
32. a) Pollinators
33. b) Cross-pollination
34. b) Tube elongation
35. a) Geitonogamy
36. b) Syngamy
37. c) Storage of nutrients
38. d) Self-incompatibility
39. a) Pollen
40. a) Produces pollen grains

SEED DEVELOPMENT AND GERMINATION

1. Which part of a seed develops into a new plant? a) Embryo b) Endosperm c) Seed coat d) Radicle

2. The process of a seed sprouting and developing into a seedling is known as: a) Pollination b) Fertilization c) Germination d) Photosynthesis

3. What is the primary function of the seed coat? a) Protection b) Absorption c) Photosynthesis d) Respiration

4. In which part of a seed does food reserves mainly get stored? a) Cotyledon b) Embryo c) Plumule d) Seed coat

5. What triggers the process of seed germination? a) Light b) Oxygen c) Water d) Carbon dioxide

6. Which hormone is responsible for seed germination? a) Auxin b) Gibberellin c) Cytokinin d) Abscisic acid

7. The first structure to emerge during germination is the: a) Plumule b) Radicle c) Hypocotyl d) Epicotyl

8. What is the purpose of stratification in seed germination? a) To break seed dormancy b) To increase water absorption c) To enhance photosynthesis d) To regulate hormone production

9. True or False: Seeds of some plants require scarification to improve germination. a) True b) False

10. Which of the following factors does NOT affect seed germination? a) Temperature b) pH level c) Seed size d) Humidity

11. Which seed germination process occurs without the presence of soil? a) Epigeal germination b) Hypogeal germination c) Cryptogeal germination d) Apogeal germination

12. Which part of the seedling grows upward, against the force of gravity? a) Radicle b) Plumule c) Hypocotyl d) Cotyledon

13. What is the role of enzymes during seed germination? a) To break down complex molecules into simpler ones b) To absorb sunlight for photosynthesis c) To regulate hormone production d) To provide structural support to the seedling

14. The process of a seedling emerging from the soil is called: a) Hypocotyl elongation b) Plumule expansion c) Cotyledon development d) Hypogeal germination

15. Which of the following is NOT a method of seed dispersal? a) Wind b) Water c) Animals d) Rhizomes

16. True or False: Seeds can remain dormant for many years before germinating. a) True b) False

17. What is the primary source of energy for the developing seedling during germination? a) Sunlight b) Stored food reserves c) Soil nutrients d) Water

18. What is the purpose of the epicotyl in seed germination? a) To protect the developing embryo b) To anchor the seed in the soil c) To absorb water from the surroundings d) To form the shoot system of the seedling

19. The process of a seedling bending toward a light source is called: a) Phototropism b) Gravitropism c) Hydrotropism d) Thigmotropism

20. What happens to the seed coat during germination? a) It disintegrates b) It hardens c) It absorbs water d) It becomes photosynthetic

21. True or False: Seeds require oxygen for germination. a) True b) False

22. Which of the following is NOT a requirement for successful seed germination? a) Light b) Temperature c) Humidity d) Air pressure

23. What is the purpose of the cotyledons in a seed? a) To store food reserves b) To absorb water c) To anchor the seed in the soil d) To regulate hormone production

24. Which of the following statements about seed development is correct? a) Seeds develop from pollen grains. b) Seeds develop from ovules. c) Seeds develop from petals. d) Seeds develop from stamens.

25. What is the process of a seedling breaking through the soil surface called? a) Imbibition b) Germination c) Emergence d) Reproduction

26. True or False: Seeds can germinate in the absence of light. a) True b) False

27. What is the purpose of the radicle in seed germination? a) To protect the seedling from pathogens b) To anchor the seed in the soil c) To absorb water from the surroundings d) To form the root system of the seedling

28. Which of the following is an example of a monocot seed? a) Bean seed b) Sunflower seed c) Corn seed d) Pea seed

29. What is the term used to describe the process of a seedling bending in response to touch? a) Phototropism b) Gravitropism c) Hydrotropism d) Thigmotropism

30. True or False: Germination is a reversible process. a) True b) False

31. What is the primary purpose of the endosperm in a seed? a) To provide nutrients to the developing embryo b) To protect the embryo from external factors c) To facilitate seed dispersal d) To absorb water during germination

32. Which part of the seedling emerges first during germination? a) Cotyledon b) Plumule c) Radicle d) Hypocotyl

33. True or False: Seeds require a specific temperature range for germination. a) True b) False

34. What is the process of a seedling bending in response to gravity called? a) Phototropism b) Gravitropism c) Hydrotropism d) Thigmotropism

35. Which of the following is an example of a dicot seed? a) Wheat seed b) Coconut seed c) Apple seed d) Orchid seed

36. What is the role of the hypocotyl in seed germination? a) To protect the developing embryo b) To anchor the seed in the soil c) To absorb water from the surroundings d) To form the root system of the seedling

37. True or False: Seeds can germinate in the absence of water. a) True b) False

38. What is the purpose of the coleoptile in seed germination? a) To protect the developing embryo b) To anchor the seed in the soil c) To absorb water from the surroundings d) To form the shoot system of the seedling

39. Which of the following statements about seed dormancy is correct?
a) Seed dormancy occurs due to high water content in the seed.
b) Seed dormancy prevents germination until conditions are favorable.
c) Seed dormancy can only be broken by light exposure.

d) Seed dormancy is a result of excessive hormone production.

40. True or False: Seeds can germinate underwater. a) True b) False

41. Which of the following is NOT a factor that affects seed germination? a) Light intensity b) Soil type c) Seed maturity d) Atmospheric pressure

42. What is the process of a seed absorbing water called? a) Imbibition b) Osmosis c) Transpiration d) Absorption

43. True or False: Some seeds require a period of cold temperatures to break dormancy. a) True b) False

44. What is the term for the protective covering that surrounds the embryo and endosperm in a seed? a) Exocarp b) Pericarp c) Testa d) Endocarp

45. Which of the following is NOT a type of seed dormancy? a) Physical dormancy b) Chemical dormancy c) Photoperiodic dormancy d) Morphological dormancy

46. What is the name of the process in which a seed resumes growth after a period of dormancy? a) Revitalization b) Rejuvenation c) Resurgence d) After-ripening

47. True or False: Germination is a continuous process that occurs throughout the life of a seed. a) True b) False

48. Which of the following is the correct order of events during seed germination?
a) Imbibition, radicle emergence, cotyledon expansion, shoot elongation
b) Imbibition, cotyledon expansion, shoot elongation, radicle emergence
c) Cotyledon expansion, shoot elongation, imbibition, radicle emergence
d) Cotyledon expansion, imbibition, radicle emergence, shoot elongation

49. What is the term for the dispersal of seeds by animals through ingestion and subsequent excretion? a) Endozoochory b) Anemochory c) Hydrochory d) Barochory

50. True or False: Light is always required for seed germination. a) True b) False

51. What is the name of the protective tissue that covers the growing tip of a root during germination? a) Root cap b) Root hair c) Root meristem d) Root cortex

52. Which of the following is NOT a plant hormone involved in seed germination? a) Ethylene b) Cytokinin c) Abscisic acid d) Melatonin

53. What is the process of a seedling growing towards a source of water called? a) Phototropism b) Hydrotropism c) Geotropism d) Chemotropism

54. True or False: Seeds can remain viable for an indefinite period under proper storage conditions. a) True b) False

55. Which of the following is a characteristic of orthodox seeds? a) They can survive extreme desiccation b) They require specialized germination conditions c) They have a short lifespan d) They are primarily wind-dispersed

56. What is the purpose of the micropyle in a seed? a) To facilitate water absorption b) To provide nutrients to the embryo c) To anchor the seed in the soil d) To release hormones during germination

57. True or False: Some seeds require exposure to fire for germination. a) True b) False

58. What is the term for the phenomenon of simultaneous germination of a group of seeds when environmental conditions are favorable? a) Synchronized germination b) Collective germination c) Mass germination d) Coordinated germination

59. Which of the following is a method of breaking seed dormancy artificially? a) Cold stratification b) Ethylene exposure c) Reducing humidity d) Increasing salinity

60. True or False: Seeds contain all the necessary nutrients for the growth of a seedling. a) True b) False

61. What is the primary function of the endosperm in a seed? a) To provide energy for germination b) To protect the embryo from pathogens c) To regulate hormone production d) To anchor the seed in the soil

62. Which of the following factors does NOT influence seed viability? a) Temperature b) Oxygen concentration c) Seed size d) Seed moisture content

63. True or False: Seeds can sense and respond to environmental cues to time their germination. a) True b) False

64. What is the name of the tissue that connects the cotyledons to the embryo in a seed? a) Hypocotyl b) Endocarp c) Hilum d) Funicle

65. Which of the following is NOT a characteristic of a viable seed? a) Hardened seed coat b) Well-developed embryo c) Adequate moisture content d) Intact endosperm

66. True or False: The rate of seed germination is influenced by genetic factors. a) True b) False

67. What is the term for the phenomenon of a seedling bending towards a light source? a) Phototropism b) Geotropism c) Hydrotropism d) Thigmotropism

68. Which of the following is an example of physical dormancy in seeds? a) Impermeable seed coat b) Chemical inhibitors c) High temperature requirements d) Irregular seed shape

69. True or False: Germination can be inhibited by the presence of allelochemicals from neighboring plants. a) True b) False

70. What is the name of the process in which a seedling becomes established in the soil and develops into a mature plant? a) Seedling growth b) Seedling establishment c) Vegetative growth d) Reproductive growth

ANSWERS

1. a) Embryo
2. c) Germination
3. a) Protection
4. a) Cotyledon
5. c) Water
6. b) Gibberellin
7. b) Radicle
8. a) To break seed dormancy
9. a) True
10. c) Seed size
11. a) Epigeal germination
12. b) Plumule
13. a) To break down complex molecules into simpler ones
14. d) Hypogeal germination
15. d) Rhizomes
16. a) True
17. b) Stored food reserves
18. d) To form the shoot system of the seedling
19. a) Phototropism
20. a) It disintegrates
21. a) True
22. d) Air pressure
23. a) To store food reserves
24. b) Seeds develop from ovules.
25. c) Emergence
26. a) True
27. b) To anchor the seed in the soil
28. c) Corn seed
29. d) Thigmotropism

30. a) True

31. a) To provide nutrients to the developing embryo

32. c) Radicle

33. a) True

34. b) Gravitropism

35. c) Apple seed

36. d) To form the root system of the seedling

37. b) False

38. d) To form the shoot system of the seedling

39. b) Seed dormancy prevents germination until conditions are favorable.

40. b) False

41. d) Atmospheric pressure

42. a) Imbibition

43. a) True

44. c) Testa

45. c) Photoperiodic dormancy

46. d) After-ripening

47. b) False

48. b) Imbibition, cotyledon expansion, shoot elongation, radicle emergence

49. a) Endozoochory

50. b) False

51. a) Root cap

52. d) Melatonin

53. b) Hydrotropism

54. a) True

55. a) They can survive extreme desiccation

56. a) To facilitate water absorption

57. a) True

58. d) Coordinated germination

59. a) Cold stratification

60. b) False

61. a) To provide energy for germination

62. c) Seed size

63. a) True

64. c) Hilum

65. a) Hardened seed coat

66. a) True

67. a) Phototropism

68. a) Impermeable seed coat

69. a) True

70. b) Seedling establishment

CHAPTER 4: PLANT ECOLOGY

ECOSYSTEMS AND ECOLOGICAL INTERACTIONS

1. Which of the following is an abiotic factor in a plant ecosystem? a) Temperature b) Predation c) Competition d) Parasitism

2. Which of the following is an example of a mutualistic interaction in plant ecosystems? a) A spider capturing and feeding on an insect b) A bird preying on a small mammal c) Bees pollinating flowers d) A deer grazing on grass

3. Which of the following is an example of a pioneer species in plant succession? a) Oak tree b) Grass c) Pine tree d) Shrub

4. Which of the following best describes a trophic level in a food chain?
a) The amount of energy transferred from one organism to another
b) The position of an organism in a food chain or web
c) The total number of organisms in a given area
d) The average biomass of organisms in an ecosystem

5. What is the primary source of energy for most ecosystems on Earth? a) Sunlight b) Wind c) Geothermal heat d) Organic matter

6. In the process of photosynthesis, plants convert sunlight into: a) Oxygen and carbon dioxide b) Glucose and water c) Oxygen and water d) Glucose and oxygen

7. What is the term for the process of converting atmospheric nitrogen into a form that plants can use? a) Nitrogen fixation b) Nitrification c) Denitrification d) Ammonification

8. Which of the following is an example of a parasitic plant? a) Sunflower b) Cactus c) Orchid d) Dodder

9. Which of the following is an example of a keystone species in a plant ecosystem? a) Oak tree b) Dandelion c) Bees d) Prairie dog

10. What is the term for the process by which water moves from plant roots to the leaves and then evaporates from the leaf surface? a) Transpiration b) Respiration c) Precipitation d) Condensation

11. Which of the following is an example of an allelopathic interaction in plant ecosystems?

a) Competition between two tree species for sunlight
b) Mutualistic relationship between bees and flowers
c) Production of toxic chemicals by a plant to inhibit the growth of nearby plants
d) Predation of insects by carnivorous plants

12. Which of the following is not a primary factor influencing plant growth in an ecosystem? a) Light availability b) Soil pH c) Temperature d) Population density

13. The process by which nutrients are returned to the soil from dead organisms is called: a) Decomposition b) Photosynthesis c) Nitrogen fixation d) Evaporation

14. Which of the following is an example of an invasive plant species? a) Rose b) Lily c) Tulip d) Kudzu

15. What is the term for a group of organisms of the same species living in the same area? a) Community b) Ecosystem c) Population d) Trophic level

16. Which of the following is an example of a secondary consumer in a food chain? a) Rabbit b) Grasshopper c) Hawk d) Grass

17. The process of transferring energy from one trophic level to another in a food chain is: a) Biomagnification b) Photosynthesis c) Trophic cascade d) Trophic transfer

18. Which of the following is an example of commensalism in plant ecosystems?
a) A bee collecting nectar from a flower
b) A vine climbing up a tree for support
c) Epiphytic orchids growing on the branches of a tree
d) A carnivorous plant trapping and digesting insects

19. What is the term for the variety of different species in an ecosystem? a) Biodiversity b) Trophic level c) Succession d) Biomass

20. Which of the following is an example of a biotic factor in a plant ecosystem? a) Rainfall b) Temperature c) Disease-causing bacteria d) Soil pH

21. Which of the following is an example of a legume plant? a) Tomato b) Cabbage c) Pea d) Potato

22. The relationship between a flowering plant and the insect that pollinates it is an example of: a) Mutualism b) Parasitism c) Commensalism d) Competition

23. Which of the following is an example of an ecosystem service provided by plants? a) Air purification b) Soil erosion c) Noise pollution d) Global warming

24. The term "ecotone" refers to:
a) The study of plant ecosystems
b) The edge between two different ecosystems
c) The process of plant succession
d) The topsoil layer in a forest ecosystem

25. Which of the following is an example of a primary producer in a food chain? a) Deer b) Lion c) Grass d) Owl

26. What is the term for the loss of water vapor from plant leaves? a) Transpiration b) Evaporation c) Condensation d) Precipitation

27. Which of the following is an example of a pioneer plant species in a volcanic area? a) Moss b) Oak tree c) Pine tree d) Fern

28. What is the term for the process by which plants convert nitrates into nitrogen gas and release it back into the atmosphere?
a) Nitrification
b) Nitrogen fixation
c) Denitrification
d) Ammonification

29. Which of the following is an example of a plant adaptation to low light conditions? a) Thick waxy leaves b) Long taproots c) Large thorns d) Brightly colored flowers

30. Which of the following is an example of an omnivorous animal in a plant ecosystem? a) Rabbit b) Grasshopper c) Hawk d) Bear

31. The relationship between a parasite and its host is an example of: a) Mutualism b) Commensalism c) Parasitism d) Competition

32. Which of the following is an example of a plant defense mechanism against herbivory?
a) Production of toxic chemicals
b) Mimicking the appearance of a predator
c) Growing thorns or spines
d) All of the above

33. What is the term for the process of converting light energy into chemical energy in plants? a) Photosynthesis b) Respiration c) Decomposition d) Transpiration

34. Which of the following is an example of a plant adaptation to dry environments? a) Large leaves b) Shallow root system c) Succulent stems d) Brightly colored flowers

35. Which of the following is an example of a plant that reproduces through vegetative propagation? a) Oak tree b) Grass c) Sunflower d) Strawberry

36. The process of converting ammonia into nitrates is called: a) Nitrification b) Nitrogen fixation c) Denitrification d) Ammonification

37. Which of the following is an example of an insectivorous plant? a) Rose b) Lily c) Venus flytrap d) Sunflower

38. What is the term for the gradual change in species composition in a community over time? a) Ecotone b) Succession c) Competition d) Biomagnification

39. Which of the following is an example of a plant adaptation to cold environments? a) Large leaves b) Deep root system c) Thick bark d) Fragrant flowers

40. The relationship between a predator and its prey is an example of: a) Mutualism b) Parasitism c) Commensalism d) Predation

41. Which of the following is an example of a plant that reproduces through rhizomes? a) Daisy b) Tulip c) Bamboo d) Fern

42. What is the term for the process of converting nitrate back into atmospheric nitrogen? a) Nitrification b) Nitrogen fixation c) Denitrification d) Ammonification

43. Which of the following is an example of a carnivorous plant? a) Oak tree b) Grass c) Venus flytrap d) Sunflower

44. What is the term for the variety of different ecosystems on Earth? a) Biodiversity b) Trophic level c) Succession d) Biomass

45. Which of the following is an example of a plant adaptation to windy environments? a) Large leaves b) Shallow root system c) Flexible stems d) Fragrant flowers

46. The process by which plants release water vapor through their leaves is called: a) Transpiration b) Evaporation c) Condensation d) Precipitation

47. Which of the following is an example of a plant that reproduces through bulbs? a) Daisy b) Tulip c) Bamboo d) Fern

48. What is the term for the process of breaking down organic matter into simpler substances? a) Decomposition b) Photosynthesis c) Nitrogen fixation d) Transpiration

49. Which of the following is an example of a plant adaptation to hot environments? a) Small leaves b) Deep root system c) Thin bark d) Succulent stems

50. The term "biome" refers to:
a) The study of plant ecosystems
b) The edge between two different ecosystems
c) A large-scale community of plants and animals
d) The topsoil layer in a forest ecosystem.

51. Which of the following terms describes the symbiotic relationship in which one organism benefits while the other is unaffected? a) Mutualism b) Commensalism c) Parasitism d) Predation

52. The process of seed dispersal by attachment to animal fur or clothing is known as: a) Endozoochory b) Anemochory c) Epizoochory d) Hydrochory

53. The primary function of the stomata in plant leaves is: a) Water absorption b) Nutrient uptake c) Gas exchange d) Photosynthesis

54. Which of the following is an example of an endangered plant species? a) Sunflower b) Dandelion c) Oak tree d) Tulip

55. The relationship between ants and acacia trees, where ants protect the tree and receive food and shelter in return, is an example of: a) Predation b) Mutualism c) Parasitism d) Amensalism

56. Which of the following is an example of a plant adaptation to low nutrient environments?
a) Rapid growth rate
b) Long taproot
c) Large leaves
d) Brightly colored flowers

57. The process by which plants absorb water through their roots is called: a) Transpiration b) Respiration c) Osmosis d) Absorption

58. The term "mycorrhizae" refers to the mutually beneficial relationship between plants and: a) Insects b) Fungi c) Bacteria d) Birds

59. Which of the following is a non-renewable resource obtained from plants? a) Timber b) Water c) Oxygen d) Sunlight

60. The process of converting atmospheric carbon dioxide into organic compounds during photosynthesis is known as: a) Carbon fixation b) Carbon cycling c) Carbon sequestration d) Carbon assimilation

61. Which of the following is an example of an allelopathic plant interaction?
a) Competition for sunlight
b) Mutualistic relationship between bees and flowers
c) Production of chemicals by one plant inhibiting the growth of another
d) Predation of insects by carnivorous plants

62. The process of converting nitrates into gaseous nitrogen by denitrifying bacteria is part of the: a) Nitrogen cycle b) Water cycle c) Carbon cycle d) Oxygen cycle

63. The concept of "ecological succession" refers to:
a) The gradual change in species composition in a community over time
b) The seasonal migration of birds
c) The transfer of energy between trophic levels
d) The competition between different plant species

64. Which of the following is an example of a plant defense mechanism against herbivory?
a) Production of toxic chemicals
b) Rapid movement
c) Camouflage
d) Symbiotic relationship with predators

65. The total amount of living matter in a given area is referred to as: a) Biomass b) Biodiversity c) Trophic level d) Ecotone

66. The role of decomposers in an ecosystem is to:
a) Produce oxygen through photosynthesis
b) Provide food for primary consumers
c) Break down dead organic matter and recycle nutrients
d) Establish a balance between predator and prey populations

67. The concept of "biomagnification" refers to the process by which:
a) Energy is transferred between trophic levels in a food chain
b) Chemical pollutants increase in concentration as they move up the food chain
c) Different plant species compete for limited resources
d) Plants convert sunlight into chemical energy through photosynthesis

68. The term "benthos" refers to organisms that live: a) In the open water of a lake or ocean b) On the surface of plants c) On the ocean floor or lake bottom d) In the canopy of trees

69. The phenomenon in which two or more species evolve adaptations in response to one another is known as: a) Coevolution b) Convergent evolution c) Divergent evolution d) Parallel evolution

70. Which of the following is an example of a plant defense mechanism against pathogens?
a) Production of toxic chemicals
b) Mimicking the appearance of a predator
c) Growing thorns or spines
d) Producing sweet nectar to attract beneficial insects

71. Which of the following terms describes the initial colonization of an area devoid of vegetation?
a) Primary succession
b) Secondary succession
c) Pioneer community
d) Climax community

72. What is the term for the process of transition from one plant community to another over time?
a) Ecological equilibrium
b) Ecological succession
c) Ecological disturbance
d) Ecological facilitation

73. Which of the following is an example of a pioneer species in primary succession? a) Oak tree b) Moss c) Maple tree d) Shrub

74. In secondary succession, the process starts from: a) Bare rock b) An existing soil base c) A pond or lake d) A climax community

75. Which of the following is an example of a seral stage in succession?
a) Early successional community
b) Climax community
c) Intermediate successional community
d) Final successional community

76. What is the term for the stable, self-sustaining community that forms at the end of succession? a) Pioneer community b) Climax community c) Successional community d) Serotinous community

77. Which of the following is an example of a mutualistic interaction involving plants? a) Pollination by bees b) Parasitism by fungi c) Predation by herbivores d) Competition between plants

78. The relationship between nitrogen-fixing bacteria and legume plants is an example of: a) Commensalism b) Mutualism c) Parasitism d) Competition

79. Which of the following is an example of an allelopathic interaction between plants?
a) Competition for sunlight
b) Nitrogen fixation by bacteria
c) Production of chemicals by one plant inhibiting the growth of another
d) Pollination by insects

80. The relationship between mycorrhizal fungi and plant roots is an example of: a) Commensalism b) Mutualism c) Parasitism d) Competition

81. What is the term for the process of two or more species evolving adaptations in response to each other? a) Co-evolution b) Convergent evolution c) Divergent evolution d) Parallel evolution

82. The relationship between ants and acacia trees, where ants protect the tree and receive food and shelter in return, is an example of: a) Mutualism b) Commensalism c) Parasitism d) Predation

83. Which of the following is an example of a facilitative interaction between plants?
a) Competition for resources
b) Production of toxic chemicals by one plant inhibiting the growth of nearby plants
c) Shade tolerance of one plant allowing another to grow beneath it
d) Direct interaction between root systems

84. Which of the following terms describes a relationship in which one organism benefits while the other is harmed? a) Mutualism b) Commensalism c) Parasitism d) Predation

85. What is the term for the process by which plants provide food and shelter for ants in exchange for protection from herbivores? a) Antagonism b) Facilitation c) Commensalism d) Mutualism

86. Which of the following is an example of a competitive interaction between plants?

a) Pollination by bees
b) Nitrogen fixation by bacteria
c) Root competition for nutrients and space
d) Mutualistic relationship between plants and mycorrhizal fungi

87. What is the term for the process of two or more species sharing a limited resource and reducing its availability to each other? a) Competition b) Mutualism c) Commensalism d) Facilitation

88. Which of the following is an example of a parasitic plant? a) Rose b) Cactus c) Orchid d) Dodder

89. The relationship between bees and flowers, where bees obtain nectar and flowers are pollinated, is an example of: a) Mutualism b) Parasitism c) Commensalism d) Predation

90. Which of the following terms describes a relationship in which one organism benefits while the other is unaffected? a) Mutualism b) Commensalism c) Parasitism d) Predation

91. Which of the following is an example of a myrmecophyte, a plant species that forms a mutualistic relationship with ants? a) Oak tree b) Cactus c) Acacia tree d) Fern

92. The relationship between fungi and plant roots, where fungi provide nutrients and receive sugars in return, is known as: a) Mutualism b) Commensalism c) Parasitism d) Predation

93. Which of the following is an example of an epiphyte, a plant that grows on other plants but does not obtain nutrients from them? a) Orchid b) Sunflower c) Grass d) Maple tree

94. The process by which plants release water vapor through their leaves is called: a) Transpiration b) Respiration c) Evaporation d) Condensation

95. The relationship between a predator and its prey is an example of: a) Mutualism b) Commensalism c) Parasitism d) Predation

96. Which of the following is an example of a plant adaptation to attract pollinators?
a) Brightly colored flowers
b) Production of toxic chemicals
c) Thorns and spines
d) Rapid growth rate

97. The relationship between nitrogen-fixing bacteria and non-legume plants is an example of: a) Commensalism b) Mutualism c) Parasitism d) Competition

98. Which of the following is an example of a plant adaptation to repel herbivores? a) Production of toxic chemicals b) Brightly colored flowers c) Nectar production d) Shade tolerance

99. The relationship between a flowering plant and the insect that pollinates it is an example of: a) Mutualism b) Parasitism c) Commensalism d) Predation

100. Which of the following is an example of a plant adaptation to ensure seed dispersal? a) Production of brightly colored fruits b) Rapid growth rate c) Long taproots d) Production of toxic chemicals

ANSWERS

1. a) Temperature
2. c) Bees pollinating flowers
3. b) Grass
4. b) The position of an organism in a food chain or web
5. a) Sunlight
6. b) Glucose and water
7. a) Nitrogen fixation
8. d) Dodder
9. d) Prairie dog
10. a) Transpiration
11. c) Production of toxic chemicals by a plant to inhibit the growth of nearby plants
12. d) Population density
13. a) Decomposition
14. d) Kudzu
15. c) Population
16. c) Hawk
17. d) Trophic transfer
18. c) Epiphytic orchids growing on the branches of a tree
19. a) Biodiversity
20. c) Disease-causing bacteria
21. c) Pea
22. a) Mutualism
23. a) Air purification
24. b) The edge between two different ecosystems
25. c) Grass
26. a) Transpiration
27. a) Moss
28. c) Denitrification
29. a) Thick waxy leaves

30. d) Bear

31. c) Parasitism

32. d) All of the above

33. a) Photosynthesis

34. c) Succulent stems

35. d) Strawberry

36. a) Nitrification

37. c) Venus flytrap

38. b) Succession

39. c) Thick bark

40. d) Predation

41. c) Bamboo

42. c) Denitrification

43. c) Venus flytrap

44. a) Biodiversity

45. c) Flexible stems

46. a) Transpiration

47. b) Tulip

48. a) Decomposition

49. d) Succulent stems

50. c) A large-scale community of plants and animals

51. b) Commensalism

52. c) Epizoochory

53. c) Gas exchange

54. d) Tulip

55. b) Mutualism

56. b) Long taproot

57. d) Absorption

58. b) Fungi

59. a) Timber

60. a) Carbon fixation

61. c) Production of chemicals by one plant inhibiting the growth of another

62. a) Nitrogen cycle

63. a) The gradual change in species composition in a community over time

64. a) Production of toxic chemicals

65. a) Biomass

66. c) Break down dead organic matter and recycle nutrients

67. b) Chemical pollutants increase in concentration as they move up the food chain

68. c) On the ocean floor or lake bottom

69. a) Coevolution

70. a) Production of toxic chemicals

71. a) Primary succession

72. b) Ecological succession

73. b) Moss

74. b) An existing soil base

75. c) Intermediate successional community

76. b) Climax community

77. a) Pollination by bees

78. b) Mutualism

79. c) Production of chemicals by one plant inhibiting the growth of another

80. b) Mutualism

81. a) Coevolution

82. a) Mutualism

83. c) Shade tolerance of one plant allowing another to grow beneath it

84. c) Parasitism

85. d) Mutualism

86. c) Root competition for nutrients and space

87. a) Competition

88. d) Dodder

89. a) Mutualism

90. b) Commensalism

91. c) Acacia tree

92. a) Mutualism

93. a) Orchid

94. a) Transpiration

95. d) Predation

96. a) Brightly colored flowers

97. b) Mutualism

98. a) Production of toxic chemicals

99. a) Mutualism

100. a) Production of brightly colored fruits

PLANT ADAPTATIONS

1. Which of the following adaptations allows certain desert plants to store water? a) Thick cuticle b) Deep root system c) Fleshy stems d) Narrow leaves

2. What type of adaptation do epiphytic plants exhibit?
a) Ability to grow in saline environments
b) Adaptation to waterlogged soils
c) Ability to obtain nutrients from air and rainwater
d) Adaptation to nutrient-rich soils

3. Which of the following is an example of a plant adaptation to attract seed-dispersing animals?
a) Production of brightly colored fruits
b) Spines or thorns
c) Presence of nectar-producing glands
d) Thin, needle-like leaves

4. What adaptation allows certain plants to survive in low-light conditions? a) Large, broad leaves b) Tall, thin stems c) Chloroplasts in leaf epidermis d) Ability to change leaf angle

5. How do succulent plants adapt to arid environments?
a) By having shallow root systems
b) By having thick, water-storing leaves or stems
c) By producing large numbers of seeds
d) By developing symbiotic relationships with fungi

6. What is the purpose of spines on cacti? a) To provide shade for the plant b) To attract pollinators c) To deter herbivores from feeding on the plant d) To facilitate seed dispersal

7. What adaptation do hydrophytic plants possess to live in aquatic environments?
a) Presence of specialized air sacs
b) Ability to float on water surfaces
c) Reduced stomata on leaf surfaces
d) Formation of adventitious roots

8. Which of the following is an example of a plant adaptation to low-nutrient environments?
a) Production of nitrogen-fixing nodules
b) Rapid growth rate

c) Shallow root system
d) Formation of symbiotic relationships with bacteria or fungi

9. How do some plants adapt to fire-prone environments? a) By developing thick bark b) By producing large numbers of seeds c) By growing tall to escape fire d) By having shallow root systems

10. What adaptation do carnivorous plants have to supplement their nutrient intake?
a) Ability to produce their own sugars through photosynthesis
b) Production of large flowers to attract pollinators
c) Development of specialized leaf structures to trap and digest insects
d) Formation of symbiotic relationships with nitrogen-fixing bacteria

11. How do xerophytic plants adapt to dry conditions?
a) By having small, needle-like leaves
b) By developing deep root systems
c) By producing fleshy stems for water storage
d) By forming associations with mycorrhizal fungi

12. What adaptation allows certain plants to disperse their seeds over long distances?
a) Production of sticky fruits
b) Formation of winged or feathery structures on seeds
c) Ability to shoot seeds explosively
d) Development of thorny structures around seeds

13. What adaptation do shade-tolerant plants possess to survive in low-light environments?
a) Large, broad leaves
b) Ability to grow tall and reach sunlight
c) Formation of symbiotic relationships with insects
d) Ability to change leaf color in response to light intensity

14. How do plants adapted to high-altitude environments cope with low oxygen levels?
a) By having large, thin leaves to maximize gas exchange
b) By producing aerenchyma tissue for increased oxygen transport
c) By developing deep root systems for oxygen uptake
d) By forming symbiotic relationships with nitrogen-fixing bacteria

15. What adaptation allows certain plants to survive in nutrient-poor or acidic soils?
a) Production of toxic compounds to deter herbivores
b) Ability to form mycorrhizal associations for enhanced nutrient uptake
c) Development of deep taproots to access nutrients deeper in the soil
d) Ability to absorb nutrients through specialized leaf structures

16. How do plants in arctic environments adapt to cold temperatures?
a) By producing antifreeze compounds to protect cells from freezing
b) By developing large leaves for efficient photosynthesis
c) By forming symbiotic relationships with nitrogen-fixing bacteria
d) By having shallow root systems to minimize exposure to frozen soil

17. What adaptation allows plants in wind-exposed areas to withstand strong gusts?
a) Production of thick, waxy cuticles
b) Formation of symbiotic relationships with mycorrhizal fungi
c) Development of flexible stems or leaves
d) Ability to close stomata to minimize water loss

18. How do plants adapt to nutrient-rich soils?
a) By producing nitrogen-fixing nodules on their roots
b) By developing shallow root systems to access surface nutrients
c) By producing large numbers of flowers for enhanced pollination
d) By forming symbiotic relationships with fungi to increase nutrient absorption

19. What adaptation allows certain plants to survive in hot and arid environments?
a) Ability to store water in specialized tissues
b) Production of brightly colored flowers to attract pollinators
c) Formation of symbiotic relationships with nitrogen-fixing bacteria
d) Development of shallow root systems for rapid water uptake

20. How do plants adapted to high-light environments protect their photosynthetic tissues from excessive light?
a) By producing UV-absorbing compounds in their leaves
b) By closing stomata during peak sunlight hours
c) By developing thick, leathery leaves for shade
d) By forming symbiotic relationships with mycorrhizal fungi

21. How do plants with needle-like leaves adapt to their environment?
a) Reduce water loss through transpiration
b) Enhance photosynthesis in low light conditions
c) Attract pollinators with their shape
d) Store nutrients for long periods

22. Which of the following adaptations allows plants to survive in nutrient-poor environments?
a) Symbiotic relationships with mycorrhizal fungi
b) Production of toxic compounds to deter herbivores
c) Thick cuticles to minimize water loss

d) Brightly colored flowers to attract pollinators

23. What is the advantage of plants with large surface area-to-volume ratios?
a) Efficient gas exchange
b) Enhanced water uptake
c) Defense against herbivores
d) Increased light absorption

24. How do plants with storage organs such as bulbs and tubers adapt to their environment?
a) Efficient nutrient storage for prolonged periods
b) Rapid growth in unpredictable conditions
d) Enhanced water absorption
d) Improved wind pollination

25. Which of the following is an example of an adaptation for seed dispersal?
a) Sticky fruits that attach to animal fur
b) Production of toxic compounds to deter herbivores
c) Thick bark to protect against fire
d) Large leaves for increased photosynthesis

26. How do plants with long taproots adapt to their environment?
a) Access deep water sources
b) Repel herbivores with toxic compounds
c) Enhance pollination by attracting specific insects
d) Adapt to low light conditions

27. What is the primary advantage of plants with thorns or spines?
a) Defense against herbivores
b) Enhanced pollination
c) Efficient water absorption
d) Increased seed production

28. Which of the following adaptations allows plants to survive in extremely dry environments?
a) CAM photosynthesis
b) Rapid growth rate
c) Chloroplasts in leaf epidermis
d) Nitrogen fixation

29. How do plants with succulent leaves adapt to their environment?
a) Store water for extended periods

b) Repel herbivores with toxic compounds
c) Attract specific pollinators with scent
d) Adapt to low light conditions

30. What is the advantage of plants with wind-pollinated flowers?
a) Greater seed production
b) Efficient water absorption
c) Enhanced pollination
d) Increased nutrient uptake

31. Which of the following adaptations allows plants to survive in habitats with regular tidal flooding? a) Ability to grow on other plants
b) Production of buoyant structures
c) Ability to absorb nutrients from the air
d) Enhanced water uptake

32. How do plants with a waxy cuticle adapt to their environment?
a) Reduce water loss through transpiration
b) Maximize light absorption
c) Attract specific pollinators with scent
d) Store nutrients for extended periods

33. What is the primary advantage of plants with deep root systems?
a) Increased access to water and nutrients
b) Enhanced pollination
c) Efficient gas exchange
d) Defense against herbivores

34. Which of the following adaptations allows plants to survive in low light conditions?
a) Large leaves for increased photosynthesis
b) Chloroplasts in stem tissues
c) Ability to grow horizontally along the ground
d) Rapid growth rate

35. How do plants with specialized root structures, such as prop roots or pneumatophores, adapt to their environment?
a) Stabilize the plant in unstable soil conditions
b) Enhance water absorption
c) Repel herbivores with toxic compounds
d) Adapt to low light conditions

36. What is the advantage of plants with a clumping or clustering growth habit?

a) Enhanced wind pollination
b) Efficient nutrient absorption
c) Defense against herbivores
d) Increased seed dispersal

37. Which of the following adaptations allows plants to survive in waterlogged or flooded environments?
a) Ability to store water in specialized tissues
b) Production of buoyant structures
c) Rapid growth rate
d) Nitrogen fixation

38. How do plants with stomatal hairs adapt to their environment?
a) Reduce water loss through transpiration
b) Increase pollen production for enhanced pollination
c) Store water for extended periods
d) Repel herbivores with toxic compounds

39. What is the primary advantage of plants with CAM photosynthesis?
a) Reduced water loss
b) Enhanced pollination
c) Improved nutrient uptake
d) Increased seed production

40. Which of the following adaptations allows plants to survive in high-salt environments?
a) Ability to absorb nutrients from the air
b) Production of buoyant structures
c) Specialized root structures for salt exclusion
d) Enhanced water uptake

41. How do plants with brightly colored flowers adapt to their environment?
a) Attract specific pollinators for efficient reproduction
b) Repel herbivores with toxic compounds
c) Store water for extended periods
d) Adapt to low light conditions

42. What is the advantage of plants with modified leaves that trap and digest insects?
a) Enhanced water uptake
b) Increased seed production
c) Efficient nutrient absorption
d) Defense against herbivores

43. Which of the following adaptations allows plants to survive in fire-prone environments?
a) Rapid growth rate after fire
b) Production of toxic compounds to deter herbivores
c) Thick bark to protect against heat and flames
d) Ability to store water in specialized tissues

44. How do plants with tendrils adapt to their environment?
a) Cling to support structures for increased stability
b) Repel herbivores with toxic compounds
c) Store water for extended periods
d) Adapt to low light conditions

45. What is the primary advantage of plants with nitrogen-fixing nodules?
a) Efficient nutrient uptake
b) Enhanced pollination
c) Increased seed production
d) Defense against pathogens

46. What adaptation allows certain plants to survive in low-water environments?
a) Ability to undergo photosynthesis at night
b) Formation of succulent leaves or stems for water storage
c) Production of dense, hairy foliage for reduced water loss
d) Development of deep taproots to reach groundwater

47. How do plants adapted to nutrient-poor aquatic environments obtain essential nutrients?
a) By producing nitrogen-fixing nodules on their roots
b) By forming symbiotic relationships with mycorrhizal fungi
c) By absorbing nutrients through their specialized aquatic leaves
d) By producing floating structures that capture nutrients from the water

48. What adaptation allows certain plants to survive in heavily shaded forest understories?
a) Ability to change leaf color based on light intensity
b) Formation of symbiotic relationships with nitrogen-fixing bacteria
c) Development of large, broad leaves for efficient light capture
d) Production of compounds that deter herbivores from feeding on leaves

49. How do plants adapted to waterlogged or swampy soils obtain oxygen?
a) By forming symbiotic relationships with nitrogen-fixing bacteria
b) By producing aerenchyma tissue for increased oxygen transport
c) By developing deep taproots to access oxygen-rich layers of soil

d) By absorbing oxygen through specialized root structures

50. What adaptation allows certain plants to survive in environments with frequent fires?
a) Ability to regenerate from underground rhizomes or bulbs
b) Formation of symbiotic relationships with mycorrhizal fungi
c) Production of seeds that require exposure to high temperatures to germinate
d) Development of thorny structures to protect against fire damage

51. How do plants adapted to nutrient-poor soils obtain essential nutrients?
a) By producing nitrogen-fixing nodules on their roots
b) By forming symbiotic relationships with mycorrhizal fungi
c) By absorbing nutrients through specialized leaf structures
d) By developing shallow root systems to access surface nutrients

52. What adaptation allows certain plants to survive in habitats with high winds?
a) Formation of dense clusters to protect against wind exposure
b) Development of flexible stems or leaves to bend with the wind
c) Production of waxy cuticles to minimize water loss in windy conditions
d) Ability to close stomata during periods of strong wind to prevent desiccation

53. How do plants adapted to dry or desert environments conserve water?
a) By developing shallow root systems for rapid water uptake
b) By producing fleshy stems or leaves for water storage
c) By forming symbiotic relationships with nitrogen-fixing bacteria
d) By undergoing photosynthesis during the night to reduce water loss

54. What adaptation allows certain plants to survive in habitats with high levels of salt?
a) Ability to excrete excess salt through specialized leaf structures
b) Formation of symbiotic relationships with mycorrhizal fungi
c) Development of deep taproots to access freshwater sources
d) Production of salt-absorbing root structures

55. How do plants adapted to shade environments maximize their light capture?
a) By developing large, broad leaves for efficient light absorption
b) By producing flowers that open and close in response to light intensity
c) By forming symbiotic relationships with nitrogen-fixing bacteria
d) By producing UV-absorbing compounds to protect against excess light

ANSWERS

1. c) Fleshy stems

2. c) Ability to obtain nutrients from air and rainwater

3. a) Production of brightly colored fruits

4. d) Ability to change leaf angle

5. b) By having thick, water-storing leaves or stems

6. c) To deter herbivores from feeding on the plant

7. d) Formation of adventitious roots

8. d) Formation of symbiotic relationships with bacteria or fungi

9. b) By producing large numbers of seeds

10. c) Development of specialized leaf structures to trap and digest insects

11. a) By having small, needle-like leaves

12. b) Formation of winged or feathery structures on seeds

13. b) Ability to grow tall and reach sunlight

14. a) By producing antifreeze compounds to protect cells from freezing

15. c) Development of deep taproots to access nutrients deeper in the soil

16. a) By producing aerenchyma tissue for increased oxygen transport

17. c) Development of flexible stems or leaves

18. b) By developing shallow root systems to access surface nutrients

19. a) Ability to store water in specialized tissues

20. a) By producing UV-absorbing compounds in their leaves

21. a) Reduce water loss through transpiration

22. a) Symbiotic relationships with mycorrhizal fungi

23. a) Efficient gas exchange

24. a) Efficient nutrient storage for prolonged periods

25. a) Sticky fruits that attach to animal fur

26. a) Access deep water sources

27. a) Defense against herbivores

28. a) CAM photosynthesis

29. a) Store water for extended periods

30. c) Enhanced pollination

31. b) Production of buoyant structures

32. a) Reduce water loss through transpiration

33. a) Increased access to water and nutrients

34. b) Chloroplasts in stem tissues

35. a) Stabilize the plant in unstable soil conditions

36. d) Increased seed dispersal

37. b) Production of buoyant structures

38. a) Reduce water loss through transpiration

39. a) Reduced water loss

40. c) Specialized root structures for salt exclusion

41. a) Attract specific pollinators for efficient reproduction

42. c) Efficient nutrient absorption

43. c) Thick bark to protect against heat and flames

44. a) Cling to support structures for increased stability

45. a) Efficient nutrient uptake

46. c) Production of dense, hairy foliage for reduced water loss

47. c) By absorbing nutrients through their specialized aquatic leaves

48. c) Development of large, broad leaves for efficient light capture

49. b) By producing aerenchyma tissue for increased oxygen transport

50. a) Ability to regenerate from underground rhizomes or bulbs

51. b) By forming symbiotic relationships with mycorrhizal fungi

52. b) Development of flexible stems or leaves to bend with the wind

53. b) By producing fleshy stems or leaves for water storage

54. a) Ability to excrete excess salt through specialized leaf structures

55. a) By developing large, broad leaves for efficient light absorption

CONSERVATION AND BIODIVERSITY

1. Which of the following factors contributes to the loss of plant biodiversity? a) Climate change b) Urbanization c) Deforestation d) All of the above

2. What is the primary reason for establishing seed banks?
a) To preserve genetic diversity of plant species
b) To enhance plant growth
c) To provide food for wildlife
d) To prevent soil erosion

3. Which of the following is an example of an endangered plant species? a) Oak tree b) Rose bush c) Venus flytrap d) Dandelion

4. What is the term for the process of intentionally transferring pollen from one plant to another? a) Pollination b) Germination c) Fertilization d) Photosynthesis

5. Which of the following is a threat to plant conservation in marine ecosystems? a) Coral bleaching b) Soil erosion c) Acid rain d) Deforestation

6. What is the role of mycorrhizal fungi in plant conservation?
a) They help plants resist diseases.
b) They facilitate nutrient absorption in plants.
c) They increase water availability in arid environments.
d) They regulate the plant's growth hormones.

7. What is the main purpose of establishing botanical gardens?
a) To showcase rare and endangered plant species
b) To promote landscaping and horticulture
c) To conduct scientific research on plants
d) To provide recreational spaces for the public

8. Which of the following is a common invasive plant species? a) Sunflower b) Tulip c) Japanese knotweed d) Lavender

9. Which of the following is a plant adaptation to fire-prone environments? a) Serotiny b) Thigmotropism c) Phototropism d) Transpiration

10. What is the purpose of conducting plant population surveys?
a) To estimate the number of plant species in an area
b) To monitor changes in plant populations over time
c) To determine the average height of plants in a forest

d) To calculate the carbon sequestration potential of plants

11. Which of the following is an example of an ex-situ plant conservation method?
a) Creating wildlife reserves
b) Establishing protected areas
c) Growing plants in a greenhouse
d) Implementing sustainable logging practices

12. What is the term for the variety of plant and animal life in a particular habitat? a) Ecosystem b) Biome c) Biodiversity d) Gene pool

13. Which of the following is NOT a direct benefit of plant biodiversity?
a) Enhanced crop productivity
b) Medicinal resources
c) Improved air quality
d) Increased soil erosion

14. Which of the following plant species is considered a "living fossil"? a) Cycad b) Fern c) Sunflower d) Moss

15. What is the primary cause of habitat fragmentation? a) Agricultural practices b) Climate change c) Pollution d) Overhunting

16. Which of the following is an example of a plant conservation organization?
a) WWF (World Wildlife Fund)
b) WHO (World Health Organization)
c) NASA (National Aeronautics and Space Administration)
d) UNESCO (United Nations Educational, Scientific and Cultural Organization)

17. What is the term for the mutual relationship between a plant and a pollinator? a) Symbiosis b) Commensalism c) Parasitism d) Mutualism

18. Which of the following is a characteristic of a successful plant invader?
a) High genetic diversity
b) Slow reproductive rate
c) Limited dispersal ability
d) Adaptability to various environments

19. What is the term for the intentional release of a captive-bred organism into its natural habitat? a) Reintroduction b) Hybridization c) Gene therapy d) Photosynthesis

20. Which of the following is an example of a plant species protected under the CITES (Convention on International Trade in Endangered Species) agreement? a) Tulip b) Rose c) Orchid d) Daffodil

21. How does plant biodiversity contribute to climate change mitigation?
a) Through carbon sequestration
b) Through ozone depletion
c) Through soil erosion
d) Through deforestation

22. What is the term for the loss of plant species from a specific habitat or ecosystem? a) Extinction b) Erosion c) Fragmentation d) Succession

23. Which of the following is an example of a keystone plant species? a) Oak tree b) Moss c) Dandelion d) Sunflower

24. What is the primary purpose of implementing sustainable logging practices?
a) To minimize the negative impact on plant biodiversity
b) To increase timber production
c) To eradicate invasive plant species
d) To encourage faster forest regeneration

25. Which of the following is a plant adaptation to low-light environments? a) Phototropism b) Chlorophyll synthesis c) Etiolation d) Stomatal opening

26. What is the term for the process of converting waste plant materials into usable energy? a) Biomass conversion b) Photosynthesis c) Germination d) Fertilization

27. How does plant diversity contribute to soil health?
a) Through nutrient cycling
b) Through atmospheric carbon absorption
c) Through water purification
d) Through erosion prevention

28. Which of the following is an example of a plant species adapted to arid environments?
a) Cactus b) Fern c) Maple tree d) Pine tree

29. What is the term for the intentional removal of invasive plant species from an ecosystem?
a) Weed control
b) Transplantation
c) Vegetation mapping

d) Forest succession

30. How does the loss of pollinators impact plant biodiversity?
a) It reduces plant reproduction and genetic diversity.
b) It enhances plant growth and seed production.
c) It promotes hybridization between plant species.
d) It increases the spread of invasive plant species.

31. Which of the following is an example of a plant adaptation to high-altitude environments? a) Thigmotropism b) Etiolation c) Succulence d) Rosette growth form

32. What is the term for the protection of entire ecosystems rather than individual species?
a) Ecosystem conservation
b) Species conservation
c) Genetic conservation
d) Habitat conservation

33. Which of the following is a plant species commonly used in traditional medicine? a) Aloe vera b) Carnation c) Peony d) Marigold

34. How does plant conservation contribute to sustainable agriculture?
a) By preserving wild crop relatives for breeding purposes
b) By promoting the use of chemical fertilizers
c) By increasing land use for agriculture
d) By encouraging monoculture farming practices

35. What is the term for the long-term change in plant communities following a disturbance? a) Succession b) Adaptation c) Evolution d) Fragmentation

36. Which of the following is a threat to plant biodiversity in freshwater ecosystems? a) Overfishing b) Acid rain c) Soil erosion d) Desertification

37. What is the term for the process of converting atmospheric nitrogen into a form usable by plants?
a) Nitrogen fixation
b) Transpiration
c) Fertilization
d) Photosynthesis

38. Which of the following is a strategy for conserving plant biodiversity in urban areas? a) Rooftop gardening b) Pesticide use c) Landfill expansion d) Industrial development

39. What is the term for the deliberate breeding of plants with desirable traits? a) Selective breeding b) Genetic modification c) Hybridization d) Cloning

40. Which of the following is a plant adaptation to low-nutrient environments? a) Mycorrhizal associations b) Leaf senescence c) Flowering d) Transpiration

41. What is the primary cause of coral bleaching, which affects marine plant biodiversity? a) Ocean acidification b) Overfishing c) Pollution d) Climate change

42. Which of the following is a characteristic of a sustainable forestry practice?
a) Regulating logging activities to minimize habitat destruction
b) Clear-cutting large sections of forests for timber extraction
c) Removing all dead trees to reduce fire risk
d) Planting non-native species to increase biodiversity

43. What is the term for the process of plants losing water through their leaves? a) Transpiration b) Evaporation c) Condensation d) Precipitation

44. Which of the following is a conservation technique that focuses on protecting plant species based on their genetic diversity?
a) Genetic conservation
b) Habitat conservation
c) Ecosystem conservation
d) Species conservation

45. How does the loss of plant biodiversity impact food security?
a) It reduces the availability of wild edible plants.
b) It enhances crop productivity and yield.
c) It promotes the use of genetically modified organisms (GMOs).
d) It increases the reliance on imported food products.

46. Which of the following is an example of a plant species adapted to fire-prone environments? a) Eucalyptus b) Maple tree c) Bamboo d) Olive tree

47. What is the term for the process of a plant bending or growing towards a light source? a) Phototropism b) Geotropism c) Thigmotropism d) Hydrotropism

48. Which of the following is a strategy for preventing the spread of invasive plant species?
a) Early detection and rapid response
b) Encouraging international trade of plant species
c) Promoting monoculture farming practices

d) Increasing deforestation in affected areas

49. What is the term for the process of protecting an endangered plant species from extinction by breeding it with closely related species?
a) Hybridization
b) Genetic modification
c) Assisted migration
d) Captive breeding

50. How does the loss of plant biodiversity impact ecosystem stability?
a) It reduces the availability of food for other organisms.
b) It enhances the resistance of ecosystems to disturbances.
c) It increases the overall resilience of ecosystems.
d) It promotes the spread of invasive plant species.

ANSWERS

1. d) All of the above

2. a) To preserve genetic diversity of plant species

3. c) Venus flytrap

4. a) Pollination

5. a) Coral bleaching

6. b) They facilitate nutrient absorption in plants.

7. a) To showcase rare and endangered plant species

8. c) Japanese knotweed

9. a) Serotiny

10. b) To monitor changes in plant populations over time

11. c) Growing plants in a greenhouse

12. c) Biodiversity

13. d) Increased soil erosion

14. a) Cycad

15. a) Agricultural practices

16. a) WWF (World Wildlife Fund)

17. d) Mutualism

18. d) Adaptability to various environments

19. a) Reintroduction

20. c) Orchid

21. a) Through carbon sequestration

22. a) Extinction

23. b) Moss

24. a) To minimize the negative impact on plant biodiversity

25. c) Etiolation

26. a) Biomass conversion

27. a) Through nutrient cycling

28. a) Cactus

29. a) Weed control

30. a) It reduces plant reproduction and genetic diversity.

31. d) Rosette growth form

32. a) Ecosystem conservation

33. a) Aloe vera

34. a) By preserving wild crop relatives for breeding purposes

35. a) Succession

36. b) Acid rain

37. a) Nitrogen fixation

38. a) Rooftop gardening

39. a) Selective breeding

40. a) Mycorrhizal associations

41. d) Climate change

42. a) Regulating logging activities to minimize habitat destruction

43. a) Transpiration

44. a) Genetic conservation

45. a) It reduces the availability of wild edible plants.

46. a) Eucalyptus

47. a) Phototropism

48. a) Early detection and rapid response

49. a) Hybridization

50. a) It reduces the availability of food for other organisms.

PLANT BIOMES AND HABITATS

1. Which of the following is characterized by extreme cold temperatures, permafrost, and short growing seasons? a) Desert b) Tundra c) Forest d) Grassland

2. The desert biome is primarily characterized by: a) High annual rainfall b) Sparse vegetation c) Dense tree cover d) Long growing seasons

3. Which of the following adaptations do desert plants typically possess?
a) Large, broad leaves for increased photosynthesis
b) Deep root systems to reach underground water sources
c) Thin, waxy cuticles to reduce water loss
d) Rapid growth rates to compensate for limited resources

4. The taiga biome is primarily characterized by:
a) Extreme heat and dry conditions
b) Rich biodiversity and tall trees
c) Freezing temperatures and permafrost
d) Abundant rainfall and dense vegetation

5. Which of the following is a common adaptation found in tundra plants?
a) Shallow root systems to access deep water sources
b) Thick, waxy cuticles to retain water
c) Extensive underground rhizome networks for storage
d) Large, colorful flowers to attract pollinators

6. Which of the following is NOT a characteristic of temperate deciduous forests?
a) Four distinct seasons
b) Moderate rainfall
c) Broadleaf trees that shed their leaves in winter
d) Harsh, arid conditions

7. What type of plant community is dominated by grasses and is found in regions with moderate rainfall? a) Desert b) Tundra c) Forest d) Grassland

8. The tropical rainforest biome is characterized by:
a) Sparse vegetation and low biodiversity
b) Extremely hot temperatures and little rainfall
c) High rainfall and rich biodiversity
d) Cold temperatures and frozen soil

9. Which of the following adaptations is common in plants found in the grassland biome?

a) Deep root systems to access underground water
b) Thick bark to protect against extreme cold
c) Succulent leaves to store water
d) Epiphytic growth to capture sunlight in the understory

10. Which of the following statements about desert plants is true?
a) They have small, needle-like leaves to reduce water loss
b) They rely on deep taproots to access underground water
c) They have broad, flat leaves for increased photosynthesis
d) They are well adapted to high rainfall and low temperatures

11. The permafrost layer found in tundra regions is:
a) A layer of peat formed from decomposed plant material
b) A thick layer of clay soil
c) A layer of permanently frozen ground
d) A layer of porous sandy soil

12. Which of the following statements about the forest biome is true?
a) It has a low diversity of plant species
b) It is characterized by low annual rainfall
c) It experiences long, cold winters and short growing seasons
d) It is dominated by coniferous trees in some regions

13. Xerophytes are plants that:
a) Require high levels of rainfall for survival
b) Thrive in waterlogged soil conditions
c) Have adaptations for arid or desert environments
d) Grow in dense shade under forest canopies

14. Which of the following adaptations is characteristic of plants in the tundra biome?
a) Large, flat leaves for increased photosynthesis
b) Extensive root systems to absorb excess water
c) Compact, cushion-like growth forms to resist wind and cold
d) Showy flowers to attract pollinators

15. The chaparral biome is characterized by:
a) Dense forests and abundant rainfall
b) Sparse vegetation and hot, dry summers
c) Harsh winters and short growing seasons
d) Tall grasses and moderate temperatures

16. Which of the following is an adaptation found in desert plants to minimize water loss?

a) Extensive root networks for water uptake
b) Large, fleshy stems for water storage
c) Thin, needle-like leaves to reduce surface area
d) Dense, waxy coatings on leaves to prevent evaporation

17. The dominant plant life form in the desert biome is: a) Trees b) Shrubs c) Grasses d) Ferns

18. Which of the following is NOT a factor contributing to the formation of deserts?
a) Proximity to mountain ranges
b) Prevailing wind patterns
c) High levels of annual rainfall
d) Cold ocean currents

19. The primary limiting factor for plant growth in the tundra biome is: a) Lack of sunlight b) Extreme heat c) Short growing seasons d) Excessive rainfall

20. Which of the following statements about plant adaptations in the forest biome is true?
a) Trees have shallow root systems due to the fertile soil.
b) Epiphytes are commonly found growing in the forest understory.
c) Broadleaf trees dominate the forest canopy to maximize sunlight absorption.
d) Forest plants have adapted to tolerate extreme arid conditions.

21. The grassland biome is characterized by:
a) Dense tree cover and abundant rainfall
b) Perennial shrubs and cold temperatures
c) Vast grassy plains and moderate rainfall
d) Rocky terrain and strong winds

22. Which of the following is NOT a characteristic of plant life in the alpine tundra?
a) Low-growing, cushion-like growth forms
b) Adaptations to strong winds and intense sunlight
c) High biodiversity and large tree species
d) Ability to tolerate freezing temperatures

23. The process by which plants lose water vapor through their leaves is called: a) Photosynthesis b) Transpiration c) Respiration d) Absorption

24. The primary limiting factor for plant growth in the desert biome is: a) Low soil fertility b) High temperatures c) Scarcity of water d) Dense vegetation cover

25. The boreal forest biome is characterized by:

a) Low temperatures and little annual rainfall
b) Dense evergreen forests and abundant wildlife
g) Drought-resistant shrubs and grasses
d) High biodiversity and tropical rainforests

26. Which of the following adaptations is common in plants found in the chaparral biome?
a) Thick bark to resist fires
b) Extensive root systems to access deep water sources
c) Large leaves to maximize photosynthesis
d) Deep taproots to anchor in rocky soils

27. The biome with the greatest diversity of plant and animal species is the: a) Tundra b) Desert c) Rainforest d) Grassland

28. Which of the following statements about the grassland biome is true?
a) It experiences low annual rainfall and drought conditions.
b) It is characterized by tall, dense forests.
c) It has a lack of herbaceous vegetation.
d) It is dominated by cacti and succulents.

29. Which of the following adaptations is characteristic of plants in the forest biome?
a) Extensive root systems to absorb water from shallow soils
b) Thick, waxy cuticles to retain water
c) Narrow leaves to minimize water loss through evaporation
d) Dense, spiky growth forms to deter herbivores

30. The process by which plants convert light energy into chemical energy is called: a) Respiration b) Photosynthesis c) Transpiration d) Pollination

31. The primary limiting factor for plant growth in the grassland biome is:
a) Short growing seasons
b) High rainfall
c) Low temperatures
d) Lack of nutrients in the soil

32. Which of the following adaptations is commonly found in plants of the temperate deciduous forest?
a) Deep taproots to access underground water
b) Thick, waxy cuticles to reduce water loss
c) Broad, flat leaves for increased photosynthesis
d) Thick, insulating bark to protect against cold temperatures

33. The steppe biome is characterized by:
a) Hot, arid conditions and sparse vegetation
b) Cold temperatures and extensive permafrost
c) Lush rainforests and high annual rainfall
d) Steep mountain slopes and alpine vegetation

34. The dominant plant life form in the forest biome is: a) Cacti b) Shrubs c) Trees d) Mosses

35. Which of the following adaptations is common in plants found in the desert biome?
a) Extensive root systems to access deep water sources
b) Broad, flat leaves to maximize sunlight absorption
c) Thorns and spines to deter herbivores
d) Epiphytic growth to capture sunlight in the understory

36. The primary limiting factor for plant growth in the boreal forest biome is:
a) High temperatures
b) Excessive rainfall
c) Lack of sunlight
d) Cold temperatures and short growing seasons

37. Which of the following statements about plant adaptations in the tundra biome is true?
a) Trees dominate the landscape due to their ability to survive in cold temperatures.
b) Plants have shallow root systems due to the rocky, permafrost soils.
c) The tundra biome has a rich diversity of flowering plants.
d) Tundra plants have adapted to withstand frequent wildfires.

38. The primary limiting factor for plant growth in the forest biome is: a) Low temperatures b) Excessive rainfall c) Lack of sunlight d) Poor soil fertility

39. Which of the following adaptations is NOT commonly found in plants of the desert biome?
a) Thick, fleshy stems for water storage
b) Small, narrow leaves to reduce water loss
c) Extensive root systems to access underground water
d) Long taproots to reach deep water sources

40. The steppe biome is commonly found in:
a) Polar regions near the North and South Poles
b) Coastal areas with high humidity and rainfall
c) Dry, continental interiors with moderate temperatures
d) Mountainous regions with rugged terrain

41. Which of the following statements about plant adaptations in the boreal forest biome is true?
a) Coniferous trees dominate the landscape due to their ability to tolerate extreme heat.
b) Plants have shallow root systems to access deep water sources.
c) The boreal forest biome has a low diversity of plant species.
d) Boreal forest plants have adapted to withstand frequent droughts.

42. The dominant plant life form in the tundra biome is: a) Grasses b) Ferns c) Mosses and lichens d) Shrubs

43. Which of the following adaptations is NOT commonly found in plants of the grassland biome?
a) Extensive root systems to access deep water sources
b) Narrow leaves to minimize water loss
c) Tall growth forms to compete for sunlight
d) Drought-resistant mechanisms to conserve water

44. The primary limiting factor for plant growth in the chaparral biome is:
a) Low temperatures and frost
b) Excessive rainfall and flooding
c) Drought and dry conditions
d) Strong winds and erosion

45. Which of the following adaptations is common in plants found in the boreal forest biome?
a) Thick, waxy cuticles to reduce water loss
b) Large, broad leaves for increased photosynthesis
c) Extensive root systems to access deep water sources
d) Compact growth forms to resist snow accumulation

46. The dominant plant life form in the grassland biome is: a) Cacti b) Grasses c) Ferns d) Trees

47. Which of the following adaptations is commonly found in plants of the chaparral biome?
a) Shallow root systems to access shallow water sources
b) Large, fleshy leaves for water storage
c) Thin, needle-like leaves to reduce water loss
d) Thick bark to resist fires

48. The primary limiting factor for plant growth in the chaparral biome is:
a) Excessive rainfall

b) High temperatures

c) Short growing seasons

d) Drought and low water availability

49. Which of the following adaptations is NOT commonly found in plants of the boreal forest biome?

a) Thick, insulating bark to protect against cold temperatures

b) Needle-like leaves to reduce water loss

c) Shallow root systems to access surface water

d) Compact growth forms to resist snow accumulation

50. The dominant plant life form in the tundra biome is: a) Grasses b) Mosses and lichens c) Shrubs d) Trees

51. Which of the following biomes is characterized by cold, dry winters and hot, dry summers, with vegetation adapted to fire? a) Tundra b) Desert c) Forest d) Chaparral

52. The temperate grassland biome is often referred to as: a) Savanna b) Steppe c) Taiga d) Mangrove

53. Which of the following adaptations is commonly found in plants of the desert biome?

a) Broad, flat leaves for increased photosynthesis

b) Large flowers to attract pollinators

c) Shallow root systems to access shallow water sources

d) Succulent stems and leaves for water storage

54. The primary limiting factor for plant growth in the tundra biome is: a) High temperatures b) Excessive rainfall c) Lack of sunlight d) Permafrost

55. The Mediterranean climate biome is characterized by:

a) Long, cold winters and short, hot summers

b) Mild, wet winters and hot, dry summers

c) Consistent rainfall throughout the year

d) Freezing temperatures and heavy snowfall

56. Which of the following adaptations is commonly found in plants of the forest biome?

a) Deep taproots to access underground water sources

b) Small, needle-like leaves to reduce water loss

c) Extensive root systems to anchor in rocky soils

d) Large, fleshy stems for water storage

57. The dominant plant life form in the alpine tundra biome is: a) Grasses b) Shrubs c) Lichens d) Mosses

58. The primary limiting factor for plant growth in the chaparral biome is:
a) High temperatures and drought
b) Excessive rainfall and flooding
c) Strong winds and erosion
d) Short growing seasons

59. Which of the following adaptations is commonly found in plants of the temperate deciduous forest?
a) Thick, waxy cuticles to reduce water loss
b) Extensive root systems to access deep water sources
c) Compact growth forms to resist snow accumulation
d) Thin, needle-like leaves to maximize sunlight absorption

60. The primary limiting factor for plant growth in the grassland biome is: a) Short growing seasons b) High rainfall c) Low temperatures d) Lack of nutrients in the soil

61. Which of the following aquatic habitats is characterized by standing water, non-flowing, and generally smaller in size?
a) Freshwater
b) Stream and rivers
c) Lakes and ponds
d) Freshwater marshes

62. Which of the following adaptations is commonly found in plants of freshwater habitats?
a) Saltwater tolerance
b) Broad, flat leaves for increased photosynthesis
c) Stiff, needle-like leaves to minimize water loss
d) Rhizomes for anchoring in moving water

63. The intertidal zone is an example of which aquatic habitat? a) Freshwater b) Marine c) Estuaries and seashores d) Deltas

64. Which of the following adaptations is common in plants of marine habitats?
a) Floating leaves to capture sunlight
b) Extensive root systems to anchor in loose sediment
c) Succulent stems and leaves for water storage
d) Broad, flat leaves for increased buoyancy

65. Which of the following aquatic habitats is characterized by the mixing of freshwater and saltwater?
a) Freshwater
b) Stream and rivers
c) Lakes and ponds
d) Estuaries and seashores

66. Mangrove forests are commonly found in which aquatic habitat? a) Freshwater b) Stream and rivers c) Lakes and ponds d) Estuaries and seashores

67. Which of the following adaptations is commonly found in plants of estuaries and seashores?
a) Shallow root systems to access surface water
b) Large, fleshy leaves for water storage
c) Dense, waxy coatings on leaves to prevent saltwater absorption
d) Epiphytic growth to capture sunlight in the understory

68. Coral reefs are an example of which aquatic habitat? a) Freshwater b) Marine c) Lakes and ponds d) Freshwater marshes

69. Which of the following adaptations is commonly found in plants of freshwater marshes?
a) Floating leaves to capture sunlight
b) Extensive root systems to anchor in loose sediment
c) Succulent stems and leaves for water storage
d) Broad, flat leaves for increased buoyancy

70. The open ocean is an example of which aquatic habitat? a) Freshwater b) Marine c) Lakes and ponds d) Freshwater marshes

71. Which of the following adaptations is commonly found in plants of deltas?
a) Shallow root systems to access surface water
b) Large, fleshy leaves for water storage
c) Dense, waxy coatings on leaves to prevent saltwater absorption
d) Epiphytic growth to capture sunlight in the understory

72. Which of the following aquatic habitats is characterized by fast-moving water and a continuous flow?
a) Freshwater
b) Stream and rivers
c) Lakes and ponds
d) Freshwater marshes

73. Which of the following adaptations is common in plants of stream and rivers?
a) Floating leaves to capture sunlight
b) Extensive root systems to anchor in loose sediment
c) Succulent stems and leaves for water storage
d) Broad, flat leaves for increased buoyancy

74. Coral reefs are highly diverse ecosystems primarily composed of: a) Seagrasses b) Algae c) Corals d) Mangroves

75. Which of the following adaptations is commonly found in plants of marine habitats to withstand wave action?
a) Stiff, needle-like leaves to minimize water loss
b) Extensive root systems to anchor in loose sediment
c) Floating leaves to capture sunlight
d) Epiphytic growth to capture sunlight in the understory

76. Which of the following aquatic habitats is characterized by low salinity, such as in rivers and streams?
a) Freshwater
b) Stream and rivers
c) Lakes and ponds
d) Freshwater marshes

77. Seagrass meadows are commonly found in which aquatic habitat? a) Freshwater b) Stream and rivers c) Lakes and ponds d) Marine

78. Which of the following adaptations is commonly found in plants of freshwater habitats to survive in low light conditions?
a) Broad, flat leaves for increased photosynthesis
b) Stiff, needle-like leaves to minimize water loss
c) Rhizomes for anchoring in moving water
d) Floating leaves to capture sunlight

79. Which of the following aquatic habitats is characterized by still, often stagnant water and abundant vegetation?
a) Freshwater
b) Stream and rivers
c) Lakes and ponds
d) Freshwater marshes

80. Which of the following adaptations is common in plants of lakes and ponds?
a) Saltwater tolerance

b) Broad, flat leaves for increased photosynthesis
c) Stiff, needle-like leaves to minimize water loss
d) Rhizomes for anchoring in moving water

81. The Great Barrier Reef is an example of which aquatic habitat? a) Freshwater b) Marine c) Lakes and ponds d) Freshwater marshes

82. Which of the following adaptations is commonly found in plants of marine habitats to cope with high salinity?
a) Floating leaves to capture sunlight
d) Extensive root systems to anchor in loose sediment
e) Succulent stems and leaves for water storage
d) Dense, waxy coatings on leaves to prevent saltwater absorption

83. Which of the following aquatic habitats is characterized by the mixing of freshwater and saltwater, with changing tidal levels?
a) Freshwater
b) Stream and rivers
c) Lakes and ponds
d) Estuaries and seashores

84. The Everglades in Florida is an example of which aquatic habitat? a) Freshwater b) Stream and rivers c) Lakes and ponds d) Freshwater marshes

85. Which of the following adaptations is commonly found in plants of estuaries and seashores to withstand wave action?
a) Shallow root systems to access surface water
b) Large, fleshy leaves for water storage
c) Dense, waxy coatings on leaves to prevent saltwater absorption
d) Epiphytic growth to capture sunlight in the understory

86. The Amazon River is an example of which aquatic habitat? a) Freshwater b) Stream and rivers c) Lakes and ponds d) Freshwater marshes

87. Which of the following adaptations is common in plants of freshwater marshes to tolerate waterlogged conditions?
a) Broad, flat leaves for increased photosynthesis
b) Stiff, needle-like leaves to minimize water loss
c) Rhizomes for anchoring in moving water
d) Floating leaves to capture sunlight

88. Kelp forests are commonly found in which aquatic habitat? a) Freshwater b) Stream and rivers c) Lakes and ponds d) Marine

89. Which of the following adaptations is commonly found in plants of marine habitats to withstand strong currents?
a) Stiff, needle-like leaves to minimize water loss
b) Extensive root systems to anchor in loose sediment
c) Floating leaves to capture sunlight
d) Epiphytic growth to capture sunlight in the understory

90. The Nile River is an example of which aquatic habitat? a) Freshwater b) Stream and rivers c) Lakes and ponds d) Freshwater marshes

91. Which of the following adaptations is common in plants of deltas to tolerate fluctuating water levels?
a) Shallow root systems to access surface water
b) Large, fleshy leaves for water storage
c) Dense, waxy coatings on leaves to prevent saltwater absorption
d) Epiphytic growth to capture sunlight in the understory

92. Which of the following aquatic habitats is characterized by a high salt concentration and large water bodies?
a) Freshwater
b) Stream and rivers
c) Lakes and ponds
d) Marine

93. Which of the following adaptations is commonly found in plants of marine habitats to withstand exposure during low tide?
a) Stiff, needle-like leaves to minimize water loss
b) Extensive root systems to anchor in loose sediment
c) Floating leaves to capture sunlight
d) Epiphytic growth to capture sunlight in the understory

94. The Dead Sea is an example of which aquatic habitat? a) Freshwater b) Stream and rivers c) Lakes and ponds d) Marine

95. Which of the following adaptations is commonly found in plants of freshwater habitats to tolerate water with low oxygen levels?
a) Broad, flat leaves for increased photosynthesis
b) Stiff, needle-like leaves to minimize water loss
c) Rhizomes for anchoring in moving water

d) Floating leaves to capture sunlight

96. Which of the following aquatic habitats is characterized by abundant floating and submerged vegetation?
a) Freshwater
b) Stream and rivers
c) Lakes and ponds
d) Freshwater marshes

97. Which of the following adaptations is common in plants of stream and rivers to withstand fast-flowing water?
a) Floating leaves to capture sunlight
b) Extensive root systems to anchor in loose sediment
c) Succulent stems and leaves for water storage
d) Broad, flat leaves for increased buoyancy

98. The Great Lakes are an example of which aquatic habitat? a) Freshwater b) Stream and rivers c) Lakes and ponds d) Freshwater marshes

99. Which of the following adaptations is commonly found in plants of lakes and ponds to survive in low light conditions?
a) Broad, flat leaves for increased photosynthesis
b) Stiff, needle-like leaves to minimize water loss
c) Rhizomes for anchoring in moving water
d) Floating leaves to capture sunlight

100. Which of the following aquatic habitats is characterized by the mixing of freshwater and saltwater, with fluctuating salinity levels?
a) Freshwater
b) Stream and rivers
c) Lakes and ponds
d) Estuaries and seashores

101. Which of the following adaptations is commonly found in plants of estuaries and seashores to cope with changing salinity levels?
a) Shallow root systems to access surface water
b) Large, fleshy leaves for water storage
c) Dense, waxy coatings on leaves to prevent saltwater absorption
d) Epiphytic growth to capture sunlight in the understory

102. Which of the following aquatic habitats is characterized by flowing water and the presence of riffles and pools? a) Freshwater b) Stream and rivers c) Lakes and ponds d) Freshwater marshes

103. Which of the following adaptations is common in plants of stream and rivers to capture nutrients from the flowing water?
a) Floating leaves to capture sunlight
b) Extensive root systems to anchor in loose sediment
c) Succulent stems and leaves for water storage
d) Broad, flat leaves for increased buoyancy

104. The Galapagos Islands are an example of which aquatic habitat? a) Freshwater b) Stream and rivers c) Lakes and ponds d) Marine

105. Which of the following adaptations is commonly found in plants of marine habitats to withstand wave action and strong currents?
a) Stiff, needle-like leaves to minimize water loss
b) Extensive root systems to anchor in loose sediment
c) Floating leaves to capture sunlight
d) Epiphytic growth to capture sunlight in the understory

106. Which of the following aquatic habitats is characterized by slow-moving or stagnant water, with abundant floating vegetation?
a) Freshwater
b) Stream and rivers
c) Lakes and ponds
d) Freshwater marshes

107. Which of the following adaptations is common in plants of lakes and ponds to capture sunlight in deeper waters?
a) Broad, flat leaves for increased photosynthesis
b) Stiff, needle-like leaves to minimize water loss
c) Rhizomes for anchoring in moving water
d) Floating leaves to capture sunlight

108. The Ganges River is an example of which aquatic habitat? a) Freshwater b) Stream and rivers c) Lakes and ponds d) Freshwater marshes

109. Which of the following adaptations is commonly found in plants of deltas to tolerate fluctuating water levels and high sedimentation?
a) Shallow root systems to access surface water
b) Large, fleshy leaves for water storage

c) Dense, waxy coatings on leaves to prevent saltwater absorption

d) Epiphytic growth to capture sunlight in the understory

110. Which of the following aquatic habitats is characterized by high salinity and diverse marine life? a) Freshwater b) Stream and rivers c) Lakes and ponds d) Marine

111. Which of the following adaptations is commonly found in plants of marine habitats to withstand exposure during low tide and high salinity?
a) Stiff, needle-like leaves to minimize water loss
b) Extensive root systems to anchor in loose sediment
c) Floating leaves to capture sunlight
d) Epiphytic growth to capture sunlight in the understory

112. The Okavango Delta is an example of which aquatic habitat? a) Freshwater b) Stream and rivers c) Lakes and ponds d) Freshwater marshes

113. Which of the following adaptations is common in plants of freshwater marshes to tolerate waterlogged conditions and high nutrient levels?
a) Broad, flat leaves for increased photosynthesis
b) Stiff, needle-like leaves to minimize water loss
c) Rhizomes for anchoring in moving water
d) Floating leaves to capture sunlight

114. Seaweed forests are commonly found in which aquatic habitat? a) Freshwater b) Stream and rivers c) Lakes and ponds d) Marine

115. Which of the following adaptations is commonly found in plants of marine habitats to withstand strong wave action and provide structural support?
a) Stiff, needle-like leaves to minimize water loss
b) Extensive root systems to anchor in loose sediment
c) Floating leaves to capture sunlight
d) Epiphytic growth to capture sunlight in the understory

116. The Mississippi River is an example of which aquatic habitat? a) Freshwater b) Stream and rivers c) Lakes and ponds d) Freshwater marshes

117. Which of the following adaptations is commonly found in plants of deltas to cope with fluctuating water levels and high sedimentation?
a) Shallow root systems to access surface water
b) Large, fleshy leaves for water storage
c) Dense, waxy coatings on leaves to prevent saltwater absorption
d) Epiphytic growth to capture sunlight in the understory

118. Which of the following aquatic habitats is characterized by the presence of floating plants and abundant birdlife?
a) Freshwater
b) Stream and rivers
c) Lakes and ponds
d) Freshwater marshes

119. Which of the following adaptations is common in plants of lakes and ponds to capture nutrients from the water column?
a) Broad, flat leaves for increased photosynthesis
b) Stiff, needle-like leaves to minimize water loss
f) Rhizomes for anchoring in moving water
d) Floating leaves to capture sunlight

120. The Great Barrier Reef is primarily composed of: a) Seagrasses b) Algae c) Corals d) Mangroves

121. What is an endangered plant species?
a) A plant species that is abundant and thriving in its natural habitat
b) A plant species that is rare and at risk of extinction
c) A plant species that has high economic value
d) A plant species that is invasive and harmful to the environment

122. Which of the following factors can contribute to the endangerment of plant species? a) Loss of habitat b) Climate change c) Overexploitation d) All of the above
123. What is the primary goal of habitat conservation?
a) To protect and preserve the genetic diversity of plant species
b) To prevent the extinction of plant species
c) To promote sustainable use of natural resources
d) All of the above

124. Which of the following actions can help conserve endangered plant species?
a) Protecting and restoring their natural habitats
b) Implementing sustainable harvesting practices
c) Establishing protected areas and reserves
d) All of the above

125. What is the role of botanical gardens in conservation efforts?
a) They provide a space for public education and awareness about endangered plants
b) They serve as repositories for preserving and propagating endangered plant species
c) They conduct research on plant conservation and habitat restoration

d) All of the above

126.What is the term used to describe the deliberate release of captive-bred individuals
    into the wild to restore or establish populations?
a) Reintroduction
b) Captive breeding
c) Habitat fragmentation
d) Invasive species

127.Which of the following organizations plays a crucial role in international plant
    conservation efforts?
a) International Union for Conservation of Nature (IUCN)
b) World Wildlife Fund (WWF)
c) United Nations Environment Programme (UNEP)
d) All of the above

128.What is the significance of seed banks in plant conservation?
a) They store and preserve seeds of endangered plant species for future use
b) They promote the exchange of plant genetic resources among scientists and researchers
c) They contribute to the restoration and recovery of degraded habitats
d) All of the above

129.What is the concept of habitat fragmentation?
a) The division of a large habitat into smaller, isolated fragments
b) The destruction of habitats due to human activities
c) The introduction of invasive species into a natural habitat
d) The loss of genetic diversity within a plant population

130.Which of the following is NOT a threat to plant habitats? a) Pollution b) Urbanization
    and habitat destruction c) Overgrazing by herbivores d) Conservation efforts

131.What is the role of citizen science in plant conservation?
a) Engaging the public in data collection and monitoring of plant species
b) Providing financial support for conservation initiatives
c) Establishing protected areas and reserves
d) Conducting research on plant genetics and breeding

132.What is the term used to describe the intentional killing, capturing, or collecting of
    plant species for trade or personal use?
a) Poaching
b) Logging
c) Deforestation

d) Habitat degradation

133. Which of the following legislative measures is aimed at protecting endangered plant
   species and their habitats?
a) The Endangered Species Act
b) The Clean Air Act
c) The Renewable Energy Standards
d) The Environmental Impact Assessment

134. What is the concept of sustainable development in relation to plant conservation?
a) Meeting the needs of the present without compromising the ability of future generations
to meet their own needs
b) Maximizing economic growth at the expense of environmental protection
c) Exploiting natural resources without considering their long-term availability
d) Promoting industrialization and urbanization without considering environmental
consequences

135. What is the role of environmental education in plant conservation?
a) Raising awareness about the importance of plants and their conservation
b) Training individuals to become botanists and plant conservationists
c) Establishing protected areas and reserves
d) Conducting research on plant genetics and breeding

136. What is the term used to describe the loss of a particular plant species from a specific
   geographic area?
a) Extinction
b) Endangerment
c) Fragmentation
d) Overexploitation

137. Which of the following factors can contribute to habitat loss for plant species?
a) Urbanization and infrastructure development
b) Deforestation and land conversion for agriculture
c) Climate change and sea level rise
d) All of the above

138. What is the importance of conducting population surveys and monitoring of
   endangered plant species?
a) To assess the status and trends of population size and distribution
b) To identify and prioritize conservation actions
c) To measure the success of conservation efforts over time
d) All of the above

139. What is the concept of in situ conservation?
a) The conservation of plant species within their natural habitats
b) The conservation of plant species in controlled environments, such as botanical gardens
c) The reintroduction of captive-bred individuals into the wild
d) The conservation of plant species through the establishment of protected areas

140. Which of the following is NOT a method of ex-situ conservation?
a) Botanical gardens and arboreta
b) Habitat restoration and protection
c) Seed banks and gene banks
d) Reintroduction programs into the wild

ANSWERS

1. b) Tundra

2. b) Sparse vegetation

3. c) Thin, waxy cuticles to reduce water loss

4. b) Rich biodiversity and tall trees

5. c) Extensive underground rhizome networks for storage

6. d) Harsh, arid conditions

7. d) Grassland

8. c) High rainfall and rich biodiversity

9. a) Deep root systems to access underground water

10. a) They have small, needle-like leaves to reduce water loss

11. c) A layer of permanently frozen ground

12. d) It is dominated by coniferous trees in some regions

13. c) Have adaptations for arid or desert environments

14. c) Compact, cushion-like growth forms to resist wind and cold

15. b) Sparse vegetation and hot, dry summers

16. d) Dense, waxy coatings on leaves to prevent evaporation

17. b) Shrubs

18. c) High levels of annual rainfall

19. c) Short growing seasons

20. b) Epiphytes are commonly found growing in the forest understory.

21. c) Vast grassy plains and moderate rainfall

22. c) High biodiversity and large tree species

23. b) Transpiration

24. c) Scarcity of water

25. b) Dense evergreen forests and abundant wildlife

26. a) Thick bark to resist fires

27. c) Rainforest

28. a) It experiences low annual rainfall and drought conditions.

29. c) Narrow leaves to minimize water loss through evaporation

30. b) Photosynthesis

31. a) Short growing seasons

32. c) Broad, flat leaves for increased photosynthesis

33. a) Hot, arid conditions and sparse vegetation

34. c) Trees

35. a) Extensive root systems to access deep water sources

36. d) Cold temperatures and short growing seasons

37. b) Plants have shallow root systems due to the rocky, permafrost soils.

38. c) Lack of sunlight

39. d) Long taproots to reach deep water sources

40. c) Dry, continental interiors with moderate temperatures

41. b) Plants have shallow root systems to access deep water sources.

42. c) Mosses and lichens

43. c) Tall growth forms to compete for sunlight

44. c) Drought and dry conditions

45. a) Thick, waxy cuticles to reduce water loss

46. b) Grasses

47. d) Thick bark to resist fires

48. d) Drought and low water availability

49. c) Shallow root systems to access surface water

50. b) Mosses and lichens

51. d) Chaparral

52. b) Steppe

53. d) Succulent stems and leaves for water storage

54. d) Permafrost

55. b) Mild, wet winters and hot, dry summers

56. b) Small, needle-like leaves to reduce water loss

57. d) Mosses

58. a) High temperatures and drought

59. c) Compact growth forms to resist snow accumulation

60. c) Low temperatures

61. c) Lakes and ponds

62. b) Broad, flat leaves for increased photosynthesis

63. c) Estuaries and seashores

64. c) Succulent stems and leaves for water storage

65. d) Estuaries and seashores

66. d) Estuaries and seashores

67. c) Dense, waxy coatings on leaves to prevent saltwater absorption

68. b) Marine

69. a) Floating leaves to capture sunlight

70. b) Marine

71. c) Dense, waxy coatings on leaves to prevent saltwater absorption

72. b) Stream and rivers

73. b) Extensive root systems to anchor in loose sediment

74. c) Corals

75. b) Extensive root systems to anchor in loose sediment

76. b) Stream and rivers

77. d) Marine

78. c) Rhizomes for anchoring in moving water

79. d) Freshwater marshes

80. a) Saltwater tolerance

81. b) Marine

82. d) Dense, waxy coatings on leaves to prevent saltwater absorption

83. c) Lakes and ponds

84. d) Freshwater marshes

85. c) Dense, waxy coatings on leaves to prevent saltwater absorption

86. b) Stream and rivers

87. a) Broad, flat leaves for increased photosynthesis

88. d) Marine

89. b) Extensive root systems to anchor in loose sediment

90. a) Freshwater

91. a) Shallow root systems to access surface water

92. d) Marine

93. b) Extensive root systems to anchor in loose sediment

94. d) Marine

95. c) Rhizomes for anchoring in moving water

96. a) Freshwater

97. b) Extensive root systems to anchor in loose sediment

98. a) Freshwater

99. a) Broad, flat leaves for increased photosynthesis

100. d) Estuaries and seashores

101. c) Dense, waxy coatings on leaves to prevent saltwater absorption

102. b) Stream and rivers

103. b) Extensive root systems to anchor in loose sediment

104. b) Marine

105. b) Extensive root systems to anchor in loose sediment

106. c) Lakes and ponds

107. d) Floating leaves to capture sunlight

108. a) Freshwater

109. a) Shallow root systems to access surface water

110. d) Marine

111. a) Stiff, needle-like leaves to minimize water loss

112. a) Freshwater

113. a) Broad, flat leaves for increased photosynthesis

114. d) Marine

115. b) Extensive root systems to anchor in loose sediment

116. b) Stream and rivers

117. b) Large, fleshy leaves for water storage

118. c) Lakes and ponds

119. c) Rhizomes for anchoring in moving water

120.c) Corals

121.b) A plant species that is rare and at risk of extinction

122.d) All of the above

123.d) All of the above

124.d) All of the above

125.d) All of the above

126.a) Reintroduction

127.a) International Union for Conservation of Nature (IUCN)

128.d) All of the above

129.a) The division of a large habitat into smaller, isolated fragments

130.d) Conservation efforts

131.a) Engaging the public in data collection and monitoring of plant species

132.a) Poaching

133.a) The Endangered Species Act

134.a) Meeting the needs of the present without compromising the ability of future generations to meet their own needs

135.a) Raising awareness about the importance of plants and their conservation

136.a) Extinction

137.d) All of the above

138.d) All of the above

139.a) The conservation of plant species within their natural habitats

140.b) Habitat restoration and protection

CHAPTER 5

PLANT EVOLUTION AND HISTORY

1. Which of the following statements about the origin of plants is correct?
a) Plants evolved from animal ancestors.
b) Plants originated from fungi.
c) Plants evolved from algae.
d) Plants have always existed in their current form.

2. The process through which plants evolved the ability to convert sunlight into chemical energy is known as:
a) Photosynthesis
b) Respiration
c) Reproduction
d) Transpiration

3. What is the primary advantage of plants transitioning from an aquatic to a terrestrial habitat?
a) Access to more nutrients
b) Protection from predators
c) Ability to reproduce more efficiently
d) Increased availability of sunlight for photosynthesis

4. Which of the following plant groups is thought to be the earliest land plants? a) Mosses b) Ferns c) Conifers d) Flowering plants

5. The development of vascular tissue allowed plants to:
a) Conduct water and nutrients throughout their bodies
b) Reproduce using flowers
c) Form seeds for reproduction
d) Carry out photosynthesis

6. What is the significance of the fossil record in understanding plant evolution?
a) It provides evidence of plant evolution and the timeline of plant history.
b) It demonstrates the existence of ancient plant species that are no longer present.
c) It helps determine the evolutionary relationships among different plant groups.
d) All of the above.

7. Which of the following plant groups dominated during the Carboniferous period, forming extensive coal deposits?
a) Gymnosperms

b) Angiosperms
c) Bryophytes
d) Pteridophytes

8. Which plant group is characterized by producing seeds enclosed within a protective structure? a) Gymnosperms b) Angiosperms c) Bryophytes d) Pteridophytes

9. The development of flowers and fruits is a characteristic feature of: a) Gymnosperms b) Angiosperms c) Bryophytes d) Pteridophytes

10. What is the significance of pollen in plant reproduction?
a) It protects the developing embryo.
b) It aids in the dispersal of male gametes.
c) It provides nutrients for the developing seed.
d) It facilitates photosynthesis in the plant.

11. Which of the following is a characteristic feature of bryophytes?
a) Presence of vascular tissue
b) Production of flowers and fruits
c) Dominance of the gametophyte generation
d) Formation of seeds for reproduction

12. Which of the following plant groups includes the largest number of species? a) Gymnosperms b) Angiosperms c) Bryophytes d) Pteridophytes

13. Which of the following is an example of a gymnosperm? a) Oak tree b) Rose bush c) Pine tree d) Fern

14. The process of double fertilization is unique to: a) Gymnosperms b) Angiosperms c) Bryophytes d) Pteridophytes

15. Which plant group is characterized by a dominant sporophyte generation and a reduced gametophyte generation?
a) Gymnosperms
b) Angiosperms
c) Bryophytes
d) Pteridophytes

16. What is the significance of seeds in plant reproduction?
a) They protect and nourish the developing embryo.
b) They aid in the dispersal of the plant species.
c) They provide a means of vegetative reproduction.

d) They facilitate pollination by attracting insects.

17. Which of the following plant groups is non-vascular? a) Mosses b) Ferns c) Conifers d) Flowering plants

18. The evolution of the cuticle, stomata, and vascular tissue allowed plants to:
a) Reduce water loss and efficiently transport water and nutrients.
b) Reproduce using flowers and fruits.
c) Adapt to dry and arid environments.
d) Carry out photosynthesis in the absence of sunlight.

19. Which of the following plant groups includes the tallest trees on Earth? a) Mosses b) Ferns c) Conifers d) Flowering plants

20. Which of the following plant groups is commonly known as "seedless" plants? a) Gymnosperms b) Angiosperms c) Bryophytes d) Pteridophytes

21. The presence of xylem and phloem is a characteristic feature of: a) Gymnosperms b) Angiosperms c) Bryophytes d) Pteridophytes

22. The process of coevolution refers to:
a) The simultaneous evolution of two or more species in response to each other.
b) The evolution of plants from animal ancestors.
c) The development of seeds in plant reproduction.
d) The establishment of symbiotic relationships among plant species.

23. Which of the following plant groups is commonly referred to as "mosses"? a) Bryophytes b) Pteridophytes c) Gymnosperms d) Angiosperms

24. The first appearance of land plants in the fossil record occurred during the: a) Silurian period b) Carboniferous period c) Jurassic period d) Cretaceous period

25. The process of alternation of generations refers to:
a) The alternating phases of diploid and haploid generations in the plant life cycle.
b) The alternating generations of plants and animals.
c) The development of flowers and fruits in plants.
d) The process of seed dispersal in plant reproduction.

26. Which of the following plant groups is commonly referred to as "ferns"? a) Bryophytes b) Pteridophytes c) Gymnosperms d) Angiosperms

27. The fossilized remains of early land plants resemble: a) Algae b) Fungi c) Insects d) Fish

28. The process of natural selection played a crucial role in:
a) Shaping the evolution and adaptation of plants to different environments.
b) The development of photosynthesis in plants.
c) The establishment of symbiotic relationships among plant species.
d) The evolution of plant reproductive structures.

29. Which of the following plant groups is characterized by the absence of true roots, stems, and leaves? a) Bryophytes b) Pteridophytes c) Gymnosperms d) Angiosperms

30. The evolution of flowers and fruits in angiosperms contributed to:
a) Increased efficiency of pollination and seed dispersal.
b) The development of vascular tissue.
c) The colonization of terrestrial habitats.
d) The adaptation to arid environments.

31. The first appearance of flowering plants in the fossil record occurred during the: a) Jurassic period b) Cretaceous period c) Paleogene period d) Neogene period

32. Which of the following plant groups is commonly referred to as "conifers"? a) Bryophytes b) Pteridophytes c) Gymnosperms d) Angiosperms

33. The process of mutualism refers to:
a) The mutually beneficial relationship between plants and animals.
b) The competition among plant species for limited resources.
c) The establishment of symbiotic relationships between plants and fungi.
d) The adaptation of plants to different environmental conditions.

34. The evolution of seeds allowed plants to:
a) Protect and nourish the developing embryo.
b) Disperse to new habitats and colonize new areas.
c) Carry out photosynthesis in the absence of sunlight.
d) Adapt to arid and water-limited environments.

35. Which of the following plant groups is commonly referred to as "flowering plants"? a) Bryophytes b) Pteridophytes c) Gymnosperms d) Angiosperms

36. The first appearance of land plants marked a significant milestone in the evolution of life on Earth because:
a) It paved the way for the colonization of terrestrial habitats.

b) It led to the development of multicellular organisms.

c) It increased the atmospheric oxygen levels.

d) It initiated the formation of complex ecosystems.

37. Which of the following plant groups is commonly referred to as "gymnosperms"? a) Bryophytes b) Pteridophytes c) Gymnosperms d) Angiosperms

38. The process of germination refers to:

a) The growth of a seed into a mature plant.

b) The process of pollination in flowering plants.

c) The formation of spores in plant reproduction.

d) The establishment of symbiotic relationships among plant species.

39. Which of the following plant groups is characterized by the production of spores for reproduction? a) Bryophytes b) Pteridophytes c) Gymnosperms d) Angiosperms

40. The evolution of multicellular plants from their unicellular ancestors is thought to have occurred approximately:

a) 500 million years ago

b) 1 billion years ago

c) 2 billion years ago

d) 4 billion years ago

41. The process of meiosis is essential for plant reproduction because it:

a) Generates haploid cells for sexual reproduction.

b) Produces diploid cells for vegetative reproduction.

c) Facilitates the growth and development of plant tissues.

d) Ensures the survival of plants in harsh environmental conditions.

42. Which of the following plant groups is commonly referred to as "angiosperms"? a) Bryophytes b) Pteridophytes c) Gymnosperms d) Angiosperms

43. The evolution of the root system in plants contributed to:

a) Increased nutrient uptake and anchoring in the soil.

b) Improved efficiency of photosynthesis.

c) Protection of the plant from herbivores and pathogens.

d) Adaptation to aquatic habitats.

44. The first land plants were likely similar to modern-day: a) Mosses b) Ferns c) Conifers d) Flowering plants

45. The process of speciation refers to: a) The formation of new plant species over time.

b) The development of polyploidy in plant cells.
c) The adaptation of plants to different environmental conditions.
d) The process of seed dispersal in plant reproduction.

46. Which of the following plant groups is commonly referred to as "seed plants"? a) Bryophytes b) Pteridophytes c) Gymnosperms d) Angiosperms

47. The evolution of the leaf structure in plants allowed for:
a) Increased surface area for photosynthesis.
b) Improved water retention and reduced transpiration.
c) Enhanced reproductive structures for seed production.
d) Adaptation to arid and water-limited environments.

48. Which of the following plant groups is commonly referred to as "pteridophytes"? a) Bryophytes b) Pteridophytes c) Gymnosperms d) Angiosperms

49. The process of polyploidy refers to:
a) The duplication of chromosomes in plant cells.
b) The formation of new plant species through hybridization.
c) The development of mutualistic relationships between plants and animals.
d) The adaptation of plants to different environmental conditions.

50. The evolution of the flower structure in plants contributed to:
a) Increased efficiency of pollination and seed production.
b) Improved water and nutrient uptake.
c) Enhanced reproductive structures for spore production.
d) Adaptation to aquatic habitats.

ANSWERS

1. c) Plants evolved from algae.

2. a) Photosynthesis

3. d) Increased availability of sunlight for photosynthesis

4. a) Mosses

5. a) Conduct water and nutrients throughout their bodies

6. d) All of the above.

7. d) Pteridophytes

8. b) Angiosperms

9. b) Angiosperms

10. b) It aids in the dispersal of male gametes.

11. c) Dominance of the gametophyte generation

12. b) Angiosperms

13. c) Pine tree

14. b) Angiosperms

15. b) Angiosperms

16. a) They protect and nourish the developing embryo.

17. a) Mosses

18. a) Reduce water loss and efficiently transport water and nutrients.

19. c) Conifers

20. d) Pteridophytes

21. b) Angiosperms

22. a) The simultaneous evolution of two or more species in response to each other.

23. a) Bryophytes

24. a) Silurian period

25. a) The alternating phases of diploid and haploid generations in the plant life cycle.

26. b) Pteridophytes

27. a) Algae

28. a) Shaping the evolution and adaptation of plants to different environments.

29. a) Bryophytes

30. a) Increased efficiency of pollination and seed dispersal.

31. b) Cretaceous period

32. c) Gymnosperms

33. a) The mutually beneficial relationship between plants and animals.

34. a) Protect and nourish the developing embryo.

35. d) Angiosperms

36. a) It paved the way for the colonization of terrestrial habitats.

37. c) Gymnosperms

38. a) The growth of a seed into a mature plant.

39. b) Pteridophytes

40. a) 500 million years ago

41. a) Generates haploid cells for sexual reproduction.

42. d) Angiosperms

43. a) Increased nutrient uptake and anchoring in the soil.

44. a) Mosses

45. a) The formation of new plant species over time.

46. c) Gymnosperms

47. a) Increased surface area for photosynthesis.

48. b) Pteridophytes

49. a) The duplication of chromosomes in plant cells.

50. a) Increased efficiency of pollination and seed production.

PLANT TAXONOMY AND CLASSIFICATION

1. Which of the following is the highest taxonomic rank in plant classification? a) Family b) Kingdom c) Genus d) Species

2. The classification system used to categorize plants based on their evolutionary relationships is called:
a) Linnaean system
b) Binomial system
c) Phylogenetic system
d) Hierarchical system

3. Which of the following is not one of the traditional kingdoms in plant classification? a) Animalia b) Plantae c) Fungi d) Protista

4. How many divisions are there in the plant kingdom? a) 3 b) 4 c) 5 d) 6

5. The division Anthophyta includes plants that: a) Produce flowers b) Reproduce through spores c) Lack true roots d) Live in aquatic environments

6. The classification of plants into families is primarily based on: a) Genetic similarities b) Leaf shape and arrangement c) Reproductive structures d) Size and height

7. Which of the following is not a characteristic used to identify and classify plants? a) Leaf color b) Flower fragrance c) Stem texture d) Soil pH tolerance

8. The scientific name of a plant consists of: a) Genus and family names b) Genus and species names c) Family and species names d) Kingdom and division names

9. The correct format for writing a scientific plant name is: a) Genus species b) Species genus c) Species only d) Genus only

10. The process of assigning a plant to a particular taxonomic group is called: a) Identification b) Classification c) Nomenclature d) Taxonomy

11. Which of the following is the correct taxonomic hierarchy?
a) Kingdom, Class, Order, Phylum, Genus, Species
b) Phylum, Class, Order, Kingdom, Genus, Species
c) Kingdom, Phylum, Class, Order, Genus, Species
d) Class, Phylum, Order, Kingdom, Genus, Species

12. Which of the following plant divisions includes the conifers? a) Magnoliophyta b) Anthophyta c) Ginkgophyta d) Pinophyta

13. The scientific name for a rose is Rosa canina. What does "canina" represent in the name? a) Family name b) Genus name c) Species name d) Common name

14. The division Bryophyta includes: a) Ferns and horsetails b) Mosses and liverworts c) Flowering plants d) Algae and lichens

15. Which of the following is not a characteristic of gymnosperms?
a) Naked seeds
b) Cones as reproductive structures
c) Vascular tissue for water transport
d) Flowers for pollination

16. Which of the following is the correct order of taxonomic ranks from broadest to most specific?
a) Class, Order, Family, Genus, Species
b) Kingdom, Phylum, Class, Order, Family
c) Family, Genus, Species, Class, Order
d) Phylum, Kingdom, Order, Genus, Family

17. The plant family Solanaceae includes which of the following plants? a) Roses b) Orchids c) Daisies d) Tomatoes

18. The division Pteridophyta includes plants that reproduce through: a) Seeds b) Spores c) Rhizomes d) Bulbs

19. The correct scientific name for a common dandelion is Taraxacum officinale. What does "officinale" signify? a) Family name b) Genus name c) Species name d) Variety name

20. The classification of plants into families and genera is primarily based on:
a) Morphological characteristics
b) Geographical distribution
c) Economic value
d) Chromosomal number

21. The plant division Coniferophyta includes: a) Flowering plants b) Ferns and horsetails c) Mosses and liverworts d) Conifers

22. Which of the following is a correct example of binomial nomenclature? a) Plantus major b) Rosa canina c) Magnolia flowers d) Genus species

23. Which of the following is the largest plant family, consisting of more than 20,000 species? a) Asteraceae b) Orchidaceae c) Fabaceae d) Rosaceae

24. The division Ginkgophyta includes only one living species, known as: a) Ginkgo biloba b) Pinus sylvestris c) Pteridium aquilinum d) Picea abies

25. The correct order of taxonomic ranks from least specific to most specific is:
a) Species, Genus, Order, Family, Class
b) Species, Genus, Family, Order, Class
c) Class, Order, Family, Genus, Species
d) Class, Family, Order, Genus, Species

26. The scientific name of a plant is the same in all languages. a) True b) False

27. The classification system proposed by Carl Linnaeus is still widely used today. a) True b) False

28. The division Magnoliophyta includes: a) Conifers b) Mosses c) Ferns d) Flowering plants

29. Which of the following is not a characteristic of angiosperms?
a) Production of flowers
b) Enclosed seeds within fruits
c) Presence of true roots, stems, and leaves
d) Reproduction through spores

30. The family Rosaceae includes which of the following plants? a) Lilies b) Roses c) Cacti d) Sunflowers

31. The correct format for writing a scientific plant name includes:
a) Underlining the entire name
b) Capitalizing the species name only
c) Italicizing the entire name
d) Capitalizing the genus name only

32. The plant division Anthocerotophyta includes: a) Hornworts b) Bryophytes c) Clubmosses d) Conifers

33. The process of identifying a plant based on its physical characteristics is known as: a) Taxonomy b) Classification c) Identification d) Nomenclature

34. The correct order of taxonomic ranks from most specific to broadest is:
a) Species, Genus, Family, Order, Class
b) Species, Genus, Order, Family, Class
c) Class, Family, Order, Genus, Species
d) Class, Order, Family, Genus, Species

35. Which of the following is not a characteristic of bryophytes?
a) Lack of vascular tissue
b) Reproduction through spores
c) Presence of flowers
d) Require a moist environment for growth

36. The division Ascomycota belongs to which kingdom? a) Plantae b) Fungi c) Protista d) Animalia

37. The correct format for writing a scientific plant name in print or typed material is: a) Italicized b) Underlined c) Bolded d) Capitalized

38. The plant family Lamiaceae includes which of the following plants? a) Orchids b) Sunflowers c) Mint and lavender d) Daisies

39. Which of the following plant divisions includes the ferns? a) Magnoliophyta b) Anthophyta c) Ginkgophyta d) Pteridophyta

40. The classification system for plants is primarily based on: a) Physical appearance b) Economic importance c) Genetic relationships d) Geographic distribution

41. The division Chlorophyta includes: a) Mosses and liverworts b) Flowering plants c) Ferns and horsetails d) Green algae

42. The correct scientific name for a common oak tree is Quercus robur. What does "robur" signify? a) Family name b) Genus name c) Species name d) Variety name

43. The plant family Fabaceae is commonly known as: a) Orchid family b) Sunflower family c) Rose family d) Pea family

44. The classification of plants into divisions is primarily based on: a) Flower structure b) Leaf characteristics c) Reproductive strategies d) Growth habits

45. The division Rhodophyta includes: a) Red algae b) Ferns and horsetails c) Mosses and liverworts d) Flowering plants

46. The scientific name of a plant is the same as its common name. a) True b) False

47. The division Cycadophyta includes plants that resemble: a) Ferns b) Palms c) Mosses d) Conifers

48. The correct order of taxonomic ranks from most specific to broadest is:
a) Species, Genus, Family, Order, Class, Kingdom
b) Species, Genus, Order, Family, Class, Kingdom
c) Kingdom, Class, Order, Family, Genus, Species
d) Kingdom, Phylum, Class, Order, Family, Genus, Species

49. The family Orchidaceae is known for its: a) Carnivorous plants b) Edible fruits c) Beautiful flowers d) Medicinal properties

50. The division Bryophyta lacks: a) True roots, stems, and leaves b) Flowers and fruits c) Chlorophyll for photosynthesis d) Spores for reproduction

ANSWERS

1. b) Kingdom

2. c) Phylogenetic system

3. a) Animalia

4. d) 6

5. a) Produce flowers

6. c) Reproductive structures

7. b) Flower fragrance

8. b) Genus and species names

9. a) Genus species

10. d) Taxonomy

11. c) Kingdom, Phylum, Class, Order, Genus, Species

12. d) Pinophyta

13. c) Species name

14. b) Mosses and liverworts

15. d) Flowers for pollination

16. a) Class, Order, Family, Genus, Species

17. d) Tomatoes

18. b) Spores

19. c) Species name

20. a) Morphological characteristics

21. d) Conifers

22. b) Rosa canina

23. a) Asteraceae

24. a) Ginkgo biloba

25. c) Class, Order, Family, Genus, Species

26. b) False

27. a) True

28. d) Flowering plants

29. d) Reproduction through spores

30. b) Roses

31. a) Underlining the entire name

32. a) Hornworts

33. c) Identification

34. b) Species, Genus, Order, Family, Class

35. c) Presence of flowers

36. b) Fungi

37. a) Italicized

38. c) Mint and lavender

39. d) Pteridophyta

40. c) Genetic relationships

41. d) Green algae

42. c) Species name

43. d) Pea family

44. c) Reproductive strategies

45. a) Red algae

46. b) False

47. b) Palms

48. d) Kingdom, Phylum, Class, Order, Family, Genus, Species

49. c) Beautiful flowers

50. a) True roots, stems, and leaves

CHAPTER 6

ETHNOBOTANY

1. What is ethnobotany?
a) The study of plant biology
b) The study of plant classification
c) The study of the relationship between plants and people
d) The study of plant genetics

2. Indigenous people use plants for: a) Ornamental purposes b) Industrial purposes c) Cultural and medicinal purposes d) Environmental conservation purposes

3. Traditional medicine refers to:
a) Modern medical practices
b) Medicinal practices that originated in the last century
c) Healing practices that have been passed down through generations
d) Experimental medical treatments

4. Which of the following is an example of plant-based remedy? a) Antibiotics b) Painkillers c) Herbal tea for digestion d) Surgical procedures

5. What is the term used for the cultural significance of plants? a) Ethnomedicine b) Ethnobotany c) Ethnopharmacology d) Ethnocultural

6. Which field of study focuses on the medicinal uses of plants by indigenous communities?
a) Ethnopharmacology
b) Ethnoecology
c) Ethnolinguistics
d) Ethnohistory

7. Which of the following is NOT a reason why traditional knowledge of plants is important?
a) Preservation of cultural heritage
b) Sustainable use of plant resources
c) Development of modern pharmaceutical drugs
d) Promotion of plant conservation

8. Which term refers to the traditional knowledge passed down orally from generation to generation? a) Ethnobotany b) Oral history c) Ethnomedicine d) Folklore

9. Indigenous people often use plants for: a) Construction materials b) Transportation purposes c) Food and nutrition d) Clothing and fashion

10. What is the term used to describe the study of the relationship between plants and human societies? a) Plant ecology b) Ethnobotany c) Horticulture d) Botanical taxonomy

11. Which of the following is an example of an indigenous use of plants?
a) Using plants for biofuel production
b) Using plants for manufacturing plastics
c) Using plants for traditional ceremonies
d) Using plants for space exploration

12. Traditional knowledge of plants is often based on:
a) Scientific experiments and studies
b) Cultural beliefs and practices
c) Government regulations and policies
d) International trade agreements

13. Which term refers to the sustainable use of plant resources? a) Ethnomedicine b) Ethnopharmacology c) Ethnoconservation d) Ethnobotany

14. What is the term used to describe the transfer of traditional plant knowledge between different cultures?
a) Ethnobotany
b) Cultural diffusion
c) Ethnoecology
d) Traditional exchange

15. Which of the following is an example of a plant with cultural significance? a) Common dandelion b) Rice plant c) Oak tree d) Venus flytrap

16. Traditional medicine systems are commonly based on: a) Synthetic chemicals b) Laboratory experiments c) Empirical observations d) Clinical trials

17. What is the term used for the identification and classification of plants by indigenous communities?
a) Ethnopharmacology
b) Ethnobotany
c) Ethnolinguistics
d) Ethnobotanical taxonomy

18. Which field of study explores the spiritual and ritualistic uses of plants by indigenous communities?
a) Ethnopharmacology
b) Ethnoecology
c) Ethnobotany
d) Ethnotheology

19. Which of the following is an example of a plant used in traditional medicine? a) Pencil cactus b) Rubber tree c) Eucalyptus tree d) Venus flytrap

20. Traditional knowledge of plants is often transmitted through: a) Written records b) Academic journals c) Oral traditions d) Social media platforms

21. Indigenous communities often rely on plants for: a) Economic development b) Environmental destruction c) Cultural identity d) Modern technology advancements

22. Which term refers to the study of the medicinal properties of plants used by indigenous communities? a) Ethnobotany b) Ethnopharmacology c) Ethnolinguistics d) Ethnoecology

23. Which of the following is NOT a factor contributing to the loss of traditional plant knowledge?
a) Climate change
b) Industrialization
c) Cultural preservation efforts
d) Globalization

24. Which term refers to the practice of using plants for culinary purposes? a) Ethnopharmacology b) Ethnoecology c) Ethnobotany d) Ethnogastronomy

25. Traditional medicine often emphasizes:
a) Personalized treatment plans
b) Standardized treatment protocols
c) Surgical interventions
d) Chemotherapy treatments

26. What is the term used for the cultural beliefs and practices associated with plants? a) Ethnomedicine b) Ethnobotany c) Ethnopharmacology d) Ethnobotanical traditions

27. Which of the following is an example of a plant used in traditional ceremonies? a) Cactus b) Wheat plant c) Poison ivy d) Tomato plant

28. Which term refers to the sustainable management and conservation of plant resources?
   a) Ethnopharmacology b) Ethnobotany c) Ethnoconservation d) Ethnogastronomy

29. Traditional knowledge of plants is often obtained through:
a) Laboratory experiments
b) Indigenous elders and community members
c) Online research databases
d) Botanical gardens

30. Indigenous communities often have:
a) Limited access to modern healthcare facilities
b) Advanced medical technologies
c) Strict government regulations on plant use
d) Decreased reliance on traditional medicine

31. Which of the following is an example of a plant used for construction purposes by indigenous communities? a) Bamboo b) Coffee plant c) Orchid d) Aloe vera

32. What is the term used for the study of the ecological relationships between plants and their environment as understood by indigenous communities?
a) Ethnoecology
b) Ethnobotany
c) Ethnolinguistics
d) Ethnopharmacology

33. Traditional medicine often utilizes plant parts such as:
a) Roots, stems, and leaves
b) Synthetic chemicals and compounds
c) Microorganisms and fungi
d) Minerals and metals

34. Which term refers to the knowledge and skills related to the cultivation of plants practiced by indigenous communities? a) Ethnopharmacology b) Ethnobotany c) Ethnoagriculture d) Ethnobotanical taxonomy

35. Which of the following is an example of a plant with economic importance to indigenous communities? a) Tulip b) Palm tree c) Venus flytrap d) Moss

36. Traditional medicine is often based on:
a) Scientific research and clinical trials
b) Traditional beliefs and cultural practices
c) Government regulations and policies

d) Modern medical advancements

37. Which term refers to the study of the language and communication related to plant knowledge by indigenous communities? a) Ethnopharmacology b) Ethnobotany c) Ethnolinguistics d) Ethnoecology

38. Indigenous communities often use plants for:
a) Cultural ceremonies
b) Genetic engineering experiments
c) Urban development projects
d) Pharmaceutical manufacturing

39. Which of the following is an example of a plant used in traditional textile production? a) Palm tree b) Coffee plant c) Cotton plant d) Poison ivy

40. Traditional knowledge of plants is often threatened by:
a) Increased recognition and support
b) Strict government regulations
c) Language revitalization efforts
d) Cultural revitalization programs

41. Which of the following is an example of a plant used in traditional crafts by indigenous communities?
a) Rose
b) Wheat plant
c) Bark of the birch tree
d) Mushroom

42. Ethnobotanical knowledge often includes:
a) Traditional plant names and classifications
b) Modern agricultural practices
c) Manufacturing techniques for synthetic drugs
d) Genetic engineering experiments

43. Traditional medicine systems are often:
a) Standardized across different cultures
b) Personalized for individual patients
c) Based on experimental research
d) Focused on surgical interventions

44. Which term refers to the study of the cultural, ecological, and economic aspects of plants used by indigenous communities? a) Ethnopharmacology b) Ethnobotany c) Ethnoconservation d) Ethnogastronomy

45. Indigenous communities often use plants for: a) Space exploration b) Energy generation c) Agricultural practices d) Industrial manufacturing

46. What is the term used for the knowledge and skills related to the preparation and administration of plant-based remedies? a) Ethnopharmacology b) Ethnobotany c) Ethnopharmacy d) Ethnogastronomy

47. Which of the following is an example of a plant used in traditional food preparation? a) Tulip b) Sugar cane c) Venus flytrap d) Poison ivy

48. Traditional medicine often combines plant-based remedies with:
a) Synthetic chemicals
b) Genetic engineering techniques
c) Physical therapy sessions
d) Dietary recommendations

49. What is the term used for the process of documenting and preserving traditional plant knowledge? a) Ethnopharmacology b) Ethnobotany c) Ethnoconservation d) Ethnographic research

50. What is the term used for the process of extracting plant compounds using solvents like ethanol or water? a) Ethnobotany b) Ethnopharmacology c) Ethnolinguistics d) Ethanolysis

51. Which of the following is NOT a factor that influences the effectiveness of traditional plant-based remedies?
a) Dosage
b) Mode of administration
c) Cultural beliefs
d) Plant color

52. What is the term used for the study of the historical uses of plants by ancient civilizations? a) Ethnobotany b) Paleobotany c) Archaeobotany d) Ethnoarchaeology

53. Which of the following is an example of a plant that has both medicinal and toxic properties? a) Aloe vera b) Stinging nettle c) Lavender d) Sunflower

54. Traditional medicine often utilizes which of the following plant parts? a) Flowers b) Bark c) Seeds d) All of the above

55. What is the term used for the study of the cultural uses of fungi by indigenous communities? a) Ethnomycology b) Ethnobotany c) Ethnozoology d) Ethnopharmacology

56. Which of the following is NOT a traditional method of preparing plant-based remedies? a) Infusion b) Distillation c) Fermentation d) Sublimation

57. Which term refers to the use of plants to repel insects or pests? a) Ethnoentomology b) Ethnobotany c) Ethnozoology d) Ethnomedicine

58. Traditional knowledge of plants is often associated with which type of learning? a) Formal education b) Informal education c) Experiential learning d) Online learning

59. Which of the following is an example of a plant used for its psychoactive properties in traditional rituals? a) Rosemary b) Sage c) Peyote d) Chamomile

60. What is the term used for the traditional practice of using plant-based remedies in combination with spiritual healing techniques? a) Ethnopharmacology b) Ethnomedicine c) Shamanism d) Psychopharmacology

61. Traditional knowledge of plants is often transmitted through: a) Songs and chants b) Pictorial representations c) Dance rituals d) All of the above

62. Which of the following is an example of a plant used for its hallucinogenic properties in traditional ceremonies? a) St. John's wort b) Ginseng c) Ayahuasca d) Peppermint

63. What is the term used for the cultural practice of using plants for divination or fortune-telling? a) Ethnomedicine b) Ethnobotany c) Ethnodivination d) Ethnopsychology

64. Traditional medicine often emphasizes the concept of: a) Individualism b) Holism c) Reductionism d) Anthropocentrism

65. Which of the following is NOT a challenge faced by indigenous communities in preserving traditional plant knowledge?
a) Loss of language and oral traditions
b) Encroachment of their lands
c) Lack of interest in traditional practices
d) Overexploitation of plant resources

66. What is the term used for the exchange of traditional plant knowledge between different generations within a community?
a) Intra-generational transmission
b) Inter-generational transmission
c) Transgenerational diffusion
d) Knowledge convergence

ANSWERS

1. c) The study of the relationship between plants and people
2. c) Cultural and medicinal purposes
3. c) Healing practices that have been passed down through generations
4. c) Herbal tea for digestion
5. b) Ethnobotany
6. a) Ethnopharmacology
7. c) Development of modern pharmaceutical drugs
8. b) Oral history
9. c) Food and nutrition
10. b) Ethnobotany
11. c) Using plants for traditional ceremonies
12. b) Cultural beliefs and practices
13. c) Ethnoconservation
14. b) Cultural diffusion
15. c) Oak tree
16. c) Empirical observations
17. b) Ethnobotany
18. d) Ethnotheology
19. c) Eucalyptus tree
20. c) Oral traditions
21. c) Cultural identity
22. b) Ethnopharmacology
23. c) Cultural preservation efforts
24. d) Ethnogastronomy
25. a) Personalized treatment plans
26. b) Ethnobotany
27. a) Cactus
28. c) Ethnoconservation
29. b) Indigenous elders and community members

30. a) Limited access to modern healthcare facilities

31. a) Bamboo

32. a) Ethnoecology

33. a) Roots, stems, and leaves

34. c) Ethnoagriculture

35. b) Palm tree

36. b) Traditional beliefs and cultural practices

37. c) Ethnolinguistics

38. a) Cultural ceremonies

39. c) Cotton plant

40. b) Strict government regulations

41. c) Bark of the birch tree

42. a) Traditional plant names and classifications

43. b) Personalized for individual patients

44. b) Ethnobotany

45. c) Agricultural practices

46. c) Ethnopharmacy

47. b) Sugar cane

48. a) Synthetic chemicals

49. d) Ethnographic research

50. b) Ethnopharmacology

51. d) Plant color

52. c) Archaeobotany

53. b) Stinging nettle

54. d) All of the above

55. a) Ethnomycology

56. d) Sublimation

57. a) Ethnoentomology

58. b) Informal education

59. c) Peyote

60. c) Shamanism

61. d) All of the above

62. c) Ayahuasca

63. c) Ethnodivination

64. b) Holism

65. c) Lack of interest in traditional practices

66. b) Inter-generational transmission

ECONOMIC BOTANY

1. Which of the following is NOT a way in which plants have influenced human civilization? a) Providing food and nutrition b) Providing shelter and building materials c) Serving as a means of transportation d) Creating weather patterns

2. Which term refers to the use of plants for medicinal purposes? a) Ethnobotany b) Herbalism c) Horticulture d) Agronomy

3. Which plant is commonly used in traditional Ayurvedic medicine? a) Eucalyptus b) Ginkgo biloba c) Neem d) Aloe vera

4. Which of the following is an example of a food crop? a) Cotton b) Rubber c) Wheat d) Bamboo

5. Which agricultural practice involves the cultivation of multiple crops in the same field? a) Monocropping b) Polyculture c) Crop rotation d) Intensive farming

6. Which plant is primarily used for the production of linen? a) Flax b) Jute c) Hemp d) Sisal

7. Which plant is the primary source of natural rubber? a) Hevea brasiliensis b) Coffea arabica c) Theobroma cacao d) Nicotiana tabacum

8. Which of the following is NOT an economic use of plants? a) Paper production b) Biofuel production c) Timber production d) Plastic production

9. Which plant is commonly used as a natural insect repellent? a) Lavender b) Peppermint c) Citronella d) Rosemary

10. Which crop is primarily responsible for the production of chocolate? a) Coffee b) Cocoa c) Tea d) Sugarcane

11. Which plant is commonly used as a source of caffeine in beverages? a) Camellia sinensis b) Coffea arabica c) Citrus sinensis d) Saccharum officinarum

12. Which plant is the primary source of vanilla flavoring? a) Cinnamon b) Nutmeg c) Vanilla planifolia d) Black pepper

13. Which of the following is an example of a medicinal plant commonly used in traditional Chinese medicine? a) Ginseng b) Oregano c) Basil d) Thyme

14. Which plant is commonly used to produce the essential oil used in perfumes? a) Rose b) Jasmine c) Lavender d) Eucalyptus

15. Which of the following plants is NOT a staple food crop? a) Rice b) Maize c) Potato d) Cotton

16. Which plant is commonly used to produce tequila? a) Agave b) Sugarcane c) Barley d) Sorghum

17. Which crop is commonly used to produce bioethanol fuel? a) Soybean b) Palm oil c) Sugarcane d) Wheat

18. Which plant is commonly used in the production of teak wood? a) Teak tree b) Oak tree c) Pine tree d) Redwood tree

19. Which plant is commonly used to produce the textile fiber known as sisal? a) Sisal plant b) Hemp plant c) Jute plant d) Flax plant

20. Which plant is commonly used as a natural dye in textile industries? a) Indigo b) Turmeric c) Henna d) Saffron

21. Which plant is commonly used to produce the popular spice known as cinnamon? a) Cinnamon tree b) Clove tree c) Nutmeg tree d) Ginger tree

22. Which crop is primarily used to produce vegetable oil? a) Sunflower b) Olive c) Mustard d) Quinoa

23. Which of the following plants is NOT commonly used in traditional Ayurvedic medicine? a) Turmeric b) Ashwagandha c) Brahmi d) Peppermint

24. Which plant is commonly used to produce the essential oil used in toothpaste and mouthwash? a) Peppermint b) Eucalyptus c) Lemon balm d) Spearmint

25. Which crop is commonly used to produce ethanol for alcoholic beverages? a) Grape b) Apple c) Barley d) Wheat

26. Which plant is commonly used as a source of fiber in the textile industry? a) Cotton b) Soybean c) Coconut d) Almond

27. Which crop is commonly used to produce biodiesel fuel? a) Rapeseed b) Corn c) Barley d) Oat

28. Which plant is commonly used to produce the popular spice known as saffron? a) Saffron crocus b) Vanilla orchid c) Rosemary d) Basil

29. Which plant is commonly used to produce the popular spice known as paprika? a) Chili pepper b) Black pepper c) Bell pepper d) Cayenne pepper

30. Which crop is commonly used to produce the popular beverage known as wine? a) Grape b) Apple c) Banana d) Coconut

31. Which plant is commonly used to produce the essential oil used in aromatherapy? a) Lavender b) Lemongrass c) Patchouli d) Ylang-ylang

32. Which of the following plants is NOT commonly used in traditional Chinese medicine? a) Astragalus b) Reishi mushroom c) Ginseng d) Eucalyptus

33. Which plant is commonly used to produce the essential oil used in skincare products? a) Tea tree b) Rosemary c) Basil d) Thyme

34. Which crop is commonly used to produce the popular beverage known as coffee? a) Coffee plant b) Tea plant c) Cocoa tree d) Sugar beet

35. Which plant is commonly used to produce the essential oil used in cooking? a) Basil b) Sage c) Rosemary d) Lemongrass

36. Which of the following plants is commonly used to produce the popular beverage known as tea? a) Camellia sinensis b) Coffea arabica c) Theobroma cacao d) Citrus sinensis

37. Which plant is commonly used to produce the popular spice known as turmeric? a) Turmeric plant b) Ginger plant c) Black pepper plant d) Cumin plant

38. Which crop is commonly used to produce the popular beverage known as beer? a) Barley b) Wheat c) Corn d) Sorghum

39. Which plant is commonly used to produce the essential oil used in perfumes? a) Rose b) Jasmine c) Lavender d) Eucalyptus

40. Which of the following plants is NOT commonly used in traditional herbal medicine? a) Echinacea b) St. John's wort c) Chamomile d) Aloe vera

41. Which crop is commonly used to produce the popular beverage known as whiskey? a) Barley b) Wheat c) Corn d) Rye

42. Which plant is commonly used as a natural sweetener? a) Stevia b) Saccharin c) Aspartame d) Sucralose

43. Which of the following plants is commonly used to produce the popular beverage known as rum? a) Sugarcane b) Wheat c) Grape d) Barley

44. Which crop is commonly used to produce the popular beverage known as vodka? a) Potato b) Rice c) Rye d) Sorghum

45. Which plant is commonly used as a source of fiber in the production of paper? a) Bamboo b) Palm c) Willow d) Oak

46. Which plant is commonly used to produce the popular spice known as cumin? a) Cumin plant b) Coriander plant c) Fennel plant d) Mustard plant

47. Which crop is commonly used to produce the popular beverage known as tequila? a) Agave b) Sugarcane c) Barley d) Sorghum

48. Which plant is commonly used to produce the essential oil used in massage therapy? a) Lavender b) Rosemary c) Eucalyptus d) Peppermint

49. Which of the following plants is commonly used in the production of furniture? a) Oak b) Maple c) Birch d) Eucalyptus

50. Which plant is commonly used to produce the essential oil used in air fresheners? a) Citrus b) Pine c) Eucalyptus d) Peppermint

51. Which plant is commonly used to produce the essential oil used in the perfume known as Chanel No. 5? a) Rose b) Jasmine c) Neroli d) Patchouli

52. Which crop is commonly used to produce the popular beverage known as sake? a) Rice b) Barley c) Corn d) Rye

53. Which plant is commonly used as a natural source of blue dye? a) Indigo b) Turmeric c) Madder root d) Annatto

54. Which crop is commonly used to produce the popular condiment known as soy sauce? a) Soybean b) Barley c) Wheat d) Rye

55. Which plant is commonly used to produce the essential oil used in the fragrance industry? a) Geranium b) Lemongrass c) Peppermint d) Sandalwood

56. Which crop is commonly used to produce the popular beverage known as whiskey in Scotland? a) Barley b) Wheat c) Corn d) Rye

57. Which plant is commonly used to produce the popular spice known as saffron? a) Saffron crocus b) Turmeric c) Annatto d) Paprika

58. Which crop is commonly used to produce the popular beverage known as rum? a) Sugarcane b) Wheat c) Grape d) Barley

59. Which plant is commonly used to produce the essential oil used in the fragrance industry? a) Rose b) Jasmine c) Lavender d) Patchouli

60. Which crop is commonly used to produce the popular beverage known as tequila? a) Agave b) Sugarcane c) Barley d) Sorghum

61. Which plant is commonly used as a natural source of red dye? a) Beetroot b) Henna c) Turmeric d) Annatto

62. Which crop is commonly used to produce the popular beverage known as vodka? a) Potato b) Rice c) Rye d) Sorghum

63. Which plant is commonly used to produce the essential oil used in the flavoring of root beer? a) Sassafras b) Wintergreen c) Birch d) Clove

64. Which crop is commonly used to produce the popular beverage known as pisco? a) Grapes b) Apples c) Corn d) Wheat

65. Which plant is commonly used to produce the essential oil used in the flavoring of Earl Grey tea? a) Bergamot orange b) Lemon c) Peppermint d) Rose

ANSWERS

1. d) Creating weather patterns
2. b) Herbalism
3. c) Neem
4. c) Wheat
5. b) Polyculture
6. a) Flax
7. a) Hevea brasiliensis
8. d) Plastic production
9. c) Citronella
10. b) Cocoa
11. b) Coffea arabica
12. c) Vanilla planifolia
13. a) Ginseng
14. b) Jasmine
15. d) Cotton
16. a) Agave
17. c) Sugarcane
18. a) Teak tree
19. a) Sisal plant
20. a) Indigo
21. a) Cinnamon tree
22. a) Sunflower
23. d) Peppermint
24. a) Peppermint
25. c) Barley
26. a) Cotton
27. a) Rapeseed
28. a) Saffron crocus
29. a) Chili pepper

30. a) Grape

31. a) Lavender

32. d) Eucalyptus

33. a) Tea tree

34. a) Coffee plant

35. c) Rosemary

36. a) Camellia sinensis

37. a) Turmeric plant

38. a) Barley

39. c) Lavender

40. d) Aloe vera

41. a) Barley

42. a) Stevia

43. a) Sugarcane

44. a) Potato

45. a) Bamboo

46. a) Cumin plant

47. a) Agave

48. a) Lavender

49. a) Oak

50. b) Pine

51. b) Jasmine

52. a) Rice

53. a) Indigo

54. a) Soybean

55. d) Sandalwood

56. a) Barley

57. a) Saffron crocus

58. a) Sugarcane

59. a) Rose

60. a) Agave

61. b) Henna

62. a) Potato

63. a) Sassafras

64. a) Grapes

65. a) Bergamot orange

PLANT DISEASES AND PESTS

1. Which of the following is not a common symptom of plant disease? a) Wilting b) Yellowing of leaves c) Excessive growth d) Leaf spots

2. What are plant pathogens?
a) Insects that damage plants
b) Microorganisms that cause plant diseases
c) Chemical substances that kill plants
d) Plants that compete with other plants

3. Which of the following is a fungal plant pathogen? a) Aphids b) Nematodes c) Rhizobium d) Rust fungus

4. What is the primary mode of transmission for viral plant diseases? a) Soil b) Water c) Insects d) Fungi

5. Which of the following is a bacterial plant pathogen? a) Whiteflies b) Fusarium c) Xanthomonas d) Scale insects

6. What is a common control measure for plant diseases caused by fungi?
a) Applying insecticides
b) Removing infected plants
c) Adjusting pH levels in the soil
d) Using traps to catch insects

7. What is the term used for the practice of growing two or more crops together to reduce pest damage? a) Crop rotation b) Companion planting c) Mulching d) Hydroponics

8. Which of the following is an example of a cultural control measure for plant diseases?
a) Using chemical pesticides
b) Planting disease-resistant varieties
c) Introducing natural predators
d) Applying fungicides

9. What is the primary function of insecticides in pest management?
a) Controlling fungal diseases
b) Preventing weed growth
c) Killing insect pests
d) Improving soil fertility

10. What is the process of monitoring and identifying pests in order to make informed pest management decisions called?
a) Integrated Pest Management (IPM)
b) Organic farming
c) Biotechnology
d) Sustainable agriculture

11. Which of the following is an example of a biological control agent used in pest management?
a) Synthetic pesticides
b) Traps and lures
c) Ladybugs
d) Chemical fertilizers

12. What is the term used for the total elimination of pests from a field or crop? a) Pest control b) Pest eradication c) Pest exclusion d) Pest resistance

13. What is the role of pheromone traps in pest management?
a) They attract and kill pests
b) They provide a habitat for natural predators
c) They monitor pest populations
d) They sterilize pests to prevent reproduction

14. Which of the following is an example of an abiotic plant disease? a) Root rot b) Powdery mildew c) Citrus canker d) Sunburn

15. Which of the following is a common insect pest that attacks tomato plants? a) Whitefly b) Aphid c) Cutworm d) All of the above

16. What is the primary mode of transmission for nematode plant diseases? a) Airborne spores b) Soil-borne organisms c) Waterborne bacteria d) Seed transmission

17. Which of the following is a cultural control measure for insect pests?
a) Using chemical insecticides
b) Introducing parasitic wasps
c) Removing weeds
d) Applying fungicides

18. What is the term used for the process of selecting and breeding plants for resistance to specific diseases or pests?
a) Genetic engineering
b) Plant biotechnology

c) Plant breeding
d) Horticulture

19. Which of the following is a fungal disease commonly found in roses? a) Tomato blight b) Black spot c) Citrus canker d) Late blight

20. What is the primary mode of transmission for bacterial plant diseases? a) Airborne spores b) Soil-borne organisms c) Waterborne bacteria d) Insect vectors

21. What is the term used for a pest control method that uses a combination of different strategies to manage pests effectively?
a) Integrated Pest Management (IPM)
b) Monoculture farming
c) Insecticide resistance
d) Organic farming

22. Which of the following is an example of a chemical control measure for plant diseases?
a) Crop rotation
b) Applying fungicides
c) Using sticky traps
d) Introducing natural predators

23. What is the primary function of fungicides in pest management? a) Controlling fungal diseases b) Preventing weed growth c) Killing insect pests d) Improving soil fertility

24. Which of the following is a viral disease commonly found in cucurbits (e.g., cucumber, melon)?
a) Tobacco mosaic virus
b) Late blight
c) Black spot
d) Verticillium wilt

25. What is the term used for a plant that is naturally resistant to a particular disease or pest? a) Host plant b) Resistant variety c) Hybrid plant d) Pathogenic plant

26. Which of the following is an example of a physical control measure for plant diseases?
a) Using chemical pesticides
b) Adjusting pH levels in the soil
c) Planting disease-resistant varieties
d) Applying fungicides

27. What is the primary mode of transmission for fungal plant diseases? a) Soil b) Water c) Insects d) Airborne spores

28. Which of the following is a cultural control measure for fungal diseases? a) Using chemical fungicides b) Removing infected plant parts c) Introducing natural predators d) Applying insecticides

29. What is the term used for the practice of planting different crops in close proximity to deter pests? a) Crop rotation b) Companion planting c) Mulching d) Hydroponics

30. What is the primary function of herbicides in pest management? a) Controlling fungal diseases b) Preventing weed growth c) Killing insect pests d) Improving soil fertility

31. Which of the following is an example of a mechanical control measure for insect pests?
a) Using chemical insecticides
b) Planting disease-resistant varieties
c) Applying sticky traps
d) Introducing parasitic wasps

32. Which of the following is a common symptom of aphid infestation on plants? a) Leaf spots b) Yellowing of leaves c) Stem rot d) All of the above

33. What is the primary mode of transmission for viral plant diseases? a) Soil b) Water c) Insects d) Fungi

34. Which of the following is an example of a bacterial disease commonly found in tomatoes? a) Late blight b) Black spot c) Bacterial wilt d) Powdery mildew

35. What is the term used for the practice of rotating crops to disrupt pest life cycles and reduce disease pressure? a) Crop rotation b) Companion planting c) Mulching d) Hydroponics

36. Which of the following is a chemical control measure for insect pests? a) Using traps and lures b) Introducing natural predators c) Applying insecticides d) Removing infected plant parts

37. What is the primary function of insecticides in pest management? a) Controlling fungal diseases b) Preventing weed growth c) Killing insect pests d) Improving soil fertility

38. Which of the following is a fungal disease commonly found in wheat? a) Tomato blight b) Black spot c) Citrus canker d) Wheat rust

39. What is the term used for a plant disease that affects the roots, causing them to rot? a) Root rot b) Powdery mildew c) Crown gall d) Fusarium wilt

40. What is the primary mode of transmission for nematode plant diseases? a) Airborne spores b) Soil-borne organisms c) Waterborne bacteria d) Seed transmission

41. Which of the following is a cultural control measure for insect pests? a) Using chemical insecticides b) Introducing parasitic wasps c) Removing weeds d) Applying fungicides

42. What is the term used for the process of genetically modifying plants to enhance their resistance to pests or diseases? a) Genetic engineering b) Plant biotechnology c) Plant breeding d) Horticulture

43. Which of the following is a fungal disease commonly found in potatoes? a) Tomato blight b) Late blight c) Citrus canker d) Black spot

44. What is the primary mode of transmission for bacterial plant diseases? a) Airborne spores b) Soil-borne organisms c) Waterborne bacteria d) Insect vectors

45. What is the term used for the practice of using multiple control methods to manage pests effectively while minimizing environmental impact? a) Integrated Pest Management (IPM) b) Monoculture farming c) Insecticide resistance d) Organic farming

46. Which of the following is an example of a chemical control measure for plant diseases? a) Crop rotation b) Applying fungicides c) Using sticky traps d) Introducing natural predators

47. What is the primary function of herbicides in pest management? a) Controlling fungal diseases b) Preventing weed growth c) Killing insect pests d) Improving soil fertility

48. Which of the following is an example of a mechanical control measure for plant diseases? a) Using chemical pesticides b) Adjusting pH levels in the soil c) Removing infected plants d) Applying fungicides

49. What is the term used for a plant disease caused by a parasitic flowering plant that invades the host plant's vascular system? a) Root rot b) Powdery mildew c) Leaf spot d) Dodder infection

50. Which of the following plant diseases is caused by a viroid, a smaller infectious agent than a virus? a) Citrus canker b) Clubroot c) Potato spindle tuber disease d) Fusarium wilt

51. What is the term used for the process of introducing a beneficial microorganism to suppress the growth of plant pathogens? a) Biological control b) Genetic engineering c) Chemical sterilization d) Phytosanitary treatment

52. Which of the following is a method for detecting plant diseases that involves cutting a plant stem and observing the color and condition of the inner tissue? a) Serological testing b) DNA sequencing c) Molecular marker analysis d) Cross-sectioning

53. What is the primary mode of transmission for phytoplasma plant diseases? a) Soil b) Water c) Insects d) Seed transmission

54. Which of the following is a common method used to control insect pests in greenhouse settings? a) Fumigation b) Soil solarization c) Cultural controls d) Companion planting

55. What is the term used for a plant disease that affects the vascular system, causing the plant to wilt and die? a) Anthracnose b) Rust c) Verticillium wilt d) Downy mildew

56. Which of the following is an example of a parasitic plant that invades the host plant's tissue and extracts nutrients? a) Dodder b) Stinging nettle c) Poison ivy d) Venus flytrap

57. What is the primary mode of transmission for fungal diseases commonly found in grains, such as wheat and barley? a) Airborne spores b) Soil-borne organisms c) Waterborne bacteria d) Seed transmission

58. Which of the following is a common plant disease caused by a soil-borne fungus that affects the roots of tomato plants? a) Verticillium wilt b) Gray mold c) Bacterial canker d) White rust

59. What is the term used for a type of pest control that involves the release of sterile insects to disrupt the pest's breeding cycle? a) Sterile insect technique b) Biological control c) Insecticidal soap application d) Trapping and monitoring

60. Which of the following is a common symptom of a viral plant disease? a) Leaf spots b) Yellowing of veins c) Rust-colored lesions d) Wilting of leaves

61. What is the term used for the practice of using physical barriers, such as nets or screens, to exclude pests from reaching plants? a) Pest control b) Pest eradication c) Pest exclusion d) Pest resistance

62. Which of the following is a plant disease caused by a fungus that primarily affects cereal crops, such as wheat and rye? a) Botrytis blight b) Clubroot c) Ergot d) Root knot nematode

63. What is the primary mode of transmission for bacterial plant diseases commonly found in citrus trees? a) Airborne spores b) Soil-borne organisms c) Waterborne bacteria d) Insect vectors

64. Which of the following is a pest control method that involves the use of pheromones to disrupt the mating patterns of insects? a) Pheromone trapping b) Biological control c) Insecticidal soap application d) Integrated Pest Management (IPM)

65. What is the term used for the practice of using a combination of different control methods, such as biological, chemical, and cultural, to manage pests effectively? a) Integrated Pest Management (IPM) b) Monoculture farming c) Insecticide resistance d) Sustainable agriculture

66. Which of the following is a fungal disease commonly found in fruit trees, such as apples and pears? a) Tomato blight b) Black spot c) Citrus canker d) Late blight

67. What is the primary function of nematicides in pest management? a) Controlling fungal diseases b) Preventing weed growth c) Killing nematode pests d) Improving soil fertility
68. Which of the following is an example of a cultural control measure for fungal diseases? a) Using chemical fungicides b) Removing infected plant parts c) Introducing natural predators d) Applying insecticides

69. What is the term used for a plant disease caused by a bacterium that results in the wilting and yellowing of leaves due to clogged vascular tissue? a) Bacterial wilt b) Crown gall c) Powdery mildew d) Leaf spot

ANSWERS

1. c) Excessive growth

2. b) Microorganisms that cause plant diseases

3. d) Rust fungus

4. c) Insects

5. c) Xanthomonas

6. b) Removing infected plants

7. b) Companion planting

8. b) Planting disease-resistant varieties

9. c) Killing insect pests

10. a) Integrated Pest Management (IPM)

11. c) Ladybugs

12. b) Pest eradication

13. c) They monitor pest populations

14. d) Sunburn

15. d) All of the above

16. b) Soil-borne organisms

17. c) Removing weeds

18. c) Plant breeding

19. b) Black spot

20. d) Insect vectors

21. a) Integrated Pest Management (IPM)

22. b) Applying fungicides

23. a) Controlling fungal diseases

24. a) Tobacco mosaic virus

25. b) Resistant variety

26. c) Planting disease-resistant varieties

27. d) Airborne spores

28. b) Removing infected plant parts

29. b) Companion planting

30. b) Preventing weed growth

31. c) Applying sticky traps

32. d) All of the above

33. c) Insects

34. c) Bacterial wilt

35. a) Crop rotation

36. c) Applying insecticides

37. c) Killing insect pests

38. d) Wheat rust

39. a) Root rot

40. b) Soil-borne organisms

41. c) Removing weeds

42. a) Genetic engineering

43. b) Late blight

44. d) Insect vectors

45. a) Integrated Pest Management (IPM)

46. b) Applying fungicides

47. b) Preventing weed growth

48. c) Removing infected plants

49. d) Dodder infection

50. c) Potato spindle tuber disease

51. a) Biological control

52. d) Cross-sectioning

53. c) Insects

54. a) Fumigation

55. c) Verticillium wilt

56. a) Dodder

57. a) Airborne spores

58. a) Verticillium wilt

59. a) Sterile insect technique

60. b) Yellowing of veins

61. c) Pest exclusion

62. c) Ergot

63. d) Insect vectors

64. a) Pheromone trapping

65. a) Integrated Pest Management (IPM)

66. b) Black spot

67. c) Killing nematode pests

68. b) Removing infected plant parts

69. a) Bacterial wilt

CHAPTER 7

PLANT GENETICS AND BIOTECHNOLOGY

1. What is the name given to the study of inherited traits in plants? a) Plant taxonomy b) Plant genetics c) Plant physiology d) Plant ecology

2. Mendel's laws of inheritance apply to the transmission of traits in plants at which level? a) Cellular level b) Organism level c) Population level d) Species level

3. The passing of traits from parents to offspring is known as: a) Genetic variation b) Genetic mutation c) Genetic recombination d) Genetic inheritance

4. Which of the following is an example of a dominant trait in plants? a) Red flower color b) White flower color c) Short plant height d) Green leaf color

5. If a plant has two alleles for a particular trait, one dominant and one recessive, which phenotype will be expressed? a) Dominant phenotype b) Recessive phenotype c) Both phenotypes d) Neither phenotype

6. The study of DNA at the molecular level is called: a) Molecular genetics b) Plant physiology c) Plant taxonomy d) Plant ecology

7. The structure of DNA was first discovered by: a) James Watson and Francis Crick b) Gregor Mendel c) Rosalind Franklin d) Thomas Hunt Morgan

8. The building blocks of DNA are called: a) Genes b) Chromosomes c) Nucleotides d) Proteins

9. DNA replication occurs during which phase of the cell cycle? a) G1 phase b) S phase c) G2 phase d) M phase

10. Which of the following enzymes is responsible for unwinding the DNA double helix during replication? a) DNA polymerase b) DNA ligase c) DNA helicase d) RNA polymerase

11. What is the purpose of PCR (polymerase chain reaction) in molecular genetics? a) To amplify DNA fragments b) To synthesize RNA molecules c) To analyze protein structures d) To create genetic mutations

12. Which of the following techniques is used to transfer genes from one organism to another? a) Polymerase chain reaction b) DNA sequencing c) Genetic engineering d) Gel electrophoresis

13. What is the function of restriction enzymes in genetic engineering? a) To replicate DNA b) To cut DNA at specific sites c) To insert genes into cells d) To amplify DNA fragments

14. Which of the following is an example of a genetically modified organism (GMO)?
a) A tomato plant bred for increased fruit size
b) A cow with a naturally occurring mutation
c) A wheat plant with a disease-resistant gene inserted
d) A wildflower found in its natural habitat

15. Genetic variation in plants can arise from: a) Mutations b) Sexual reproduction c) Both mutations and sexual reproduction d) Neither mutations nor sexual reproduction

16. Which of the following is an example of a point mutation? a) Deletion of a DNA segment b) Duplication of a DNA segment c) Substitution of one base pair for another d) Inversion of a DNA segment

17. Which of the following is an example of a chromosomal mutation? a) Frameshift mutation b) Silent mutation c) Deletion of a chromosome segment d) Substitution of a single nucleotide

18. The study of how genes are expressed and regulated is known as: a) Transcription b) Translation c) Gene expression d) Genetic engineering

19. What is the purpose of a promoter region in DNA? a) To code for proteins b) To initiate DNA replication c) To start transcription of a gene d) To repair DNA damage

20. Which of the following is an example of an environmental factor that can influence gene expression in plants? a) Temperature b) Soil pH c) Light intensity d) All of the above

21. What is the role of tRNA in protein synthesis?
a) It carries the genetic code from DNA to the ribosome.
b) It carries amino acids to the ribosome for protein synthesis.
c) It helps unwind the DNA double helix during transcription.
d) It catalyzes the formation of peptide bonds between amino acids.

22. Which of the following is a technique used to analyze and compare DNA sequences?
a) DNA replication b) DNA sequencing c) DNA cloning d) DNA hybridization

23. What is the purpose of gel electrophoresis in molecular genetics? a) To amplify DNA fragments b) To cut DNA at specific sites c) To separate DNA fragments based on size d) To insert genes into cells

24. Which of the following is an application of plant tissue culture in biotechnology? a) Cloning plants b) Creating hybrid crops c) Producing genetically modified organisms d) All of the above

25. What is the purpose of the Ti plasmid in genetic engineering? a) To cut DNA at specific sites b) To insert genes into plant cells c) To amplify DNA fragments d) To create genetic mutations

26. Which of the following techniques is used to introduce foreign genes into plant cells? a) Agrobacterium-mediated transformation b) Polymerase chain reaction c) DNA sequencing d) Gel electrophoresis

27. Which of the following is an example of a transgenic plant?
a) A plant with genes from a different plant species
b) A plant with a naturally occurring mutation
c) A plant with genes from a different animal species
d) A plant found in its natural habitat

28. What is the purpose of herbicide-resistant crops in agriculture? a) To increase crop yields b) To control pests and diseases c) To reduce the use of chemical herbicides d) To improve nutrient content in crops

29. Which of the following techniques is used to detect and analyze gene expression in plants? a) Polymerase chain reaction b) DNA sequencing c) Microarray analysis d) Gel electrophoresis

30. What is the purpose of gene editing techniques, such as CRISPR-Cas9, in plant biotechnology? a) To study gene expression patterns b) To create genetic mutations c) To insert genes into plant cells d) To control gene expression

31. Which of the following is an example of a molecular marker used in plant genetics? a) DNA polymerase b) RNA polymerase c) Restriction enzyme d) AFLP (Amplified Fragment Length Polymorphism)

32. What is the role of phytohormones in plant growth and development? a) To regulate gene expression b) To provide energy for cellular processes c) To control water uptake in plants d) To absorb sunlight for photosynthesis

33. Which of the following is a natural process of transferring genetic material between plants? a) Genetic engineering b) Genetic variation c) Genetic mutation d) Gene flow

34. Which of the following techniques is used to assess genetic relatedness among plant species? a) DNA sequencing b) DNA replication c) DNA hybridization d) DNA cloning

35. Which of the following is an example of a plant breeding technique that does not involve genetic modification? a) Hybridization b) Transgenesis c) RNA interference d) CRISPR-Cas9

36. What is the function of RNA interference (RNAi) in plant biotechnology? a) To amplify DNA fragments b) To cut DNA at specific sites c) To insert genes into cells d) To regulate gene expression

37. What is the purpose of gene banks in plant conservation? a) To store and preserve plant genetic material b) To produce genetically modified organisms c) To analyze gene expression patterns d) To control gene expression

38. Which of the following is a limitation of genetic engineering in plants? a) Loss of genetic diversity b) Increased crop yields c) Reduced pesticide use d) Improved nutritional content

39. Which of the following is an example of a plant hormone involved in fruit ripening? a) Auxin b) Cytokinin c) Ethylene d) Gibberellin

40. What is the purpose of marker-assisted selection in plant breeding? a) To create genetic mutations b) To insert genes into plant cells c) To control gene expression d) To select plants with desired traits based on molecular markers

41. Which of the following is an example of a genetically modified crop that is resistant to pests? a) Bt cotton b) Wheat with increased yield c) Rice with enhanced nutritional content d) Soybean with herbicide resistance

42. What is the role of endonucleases in DNA cloning? a) To replicate DNA b) To cut DNA at specific sites c) To insert genes into cells d) To amplify DNA fragments

43. Which of the following is a technique used to analyze gene expression on a genome-wide scale? a) Polymerase chain reaction b) DNA sequencing c) Microarray analysis d) Gel electrophoresis

44. What is the purpose of biofortification in plant biotechnology? a) To study gene expression patterns b) To create genetic mutations c) To increase nutrient content in crops d) To control gene expression

45. Which of the following is an example of a somatic hybridization technique used in plant biotechnology? a) Agrobacterium-mediated transformation b) Protoplast fusion c) DNA sequencing d) Gel electrophoresis

46. What is the purpose of plant transformation in genetic engineering? a) To study gene expression patterns b) To create genetic mutations c) To insert genes into plant cells d) To control gene expression

47. Which of the following is an example of a molecular marker used to identify individuals in plant populations? a) DNA polymerase b) RNA polymerase c) Restriction enzyme d) Microsatellite marker

48. What is the function of the Cas9 enzyme in the CRISPR-Cas9 system? a) To amplify DNA fragments b) To cut DNA at specific sites c) To insert genes into cells d) To regulate gene expression

49. Which of the following is an example of a stress-tolerant genetically modified crop? a) Maize with increased yield b) Wheat with herbicide resistance c) Rice with drought tolerance d) Soybean with enhanced nutritional content

50. What is the purpose of genome editing techniques in plant biotechnology? a) To study gene expression patterns b) To create genetic mutations c) To insert genes into plant cells d) To control gene expression

51. What is the term used to describe the phenomenon in which a single gene influences multiple traits? a) Polygenic inheritance b) Pleiotropy c) Epistasis d) Codominance

52. In a plant population, a trait exhibits continuous variation. What type of inheritance pattern is most likely responsible for this? a) Mendelian inheritance b) Polygenic inheritance c) Codominance d) X-linked inheritance

53. Which of the following genetic disorders is caused by the deletion of a small portion of chromosome 15 and is characterized by intellectual disabilities, behavioral problems, and a happy demeanor? a) Turner syndrome b) Down syndrome c) Prader-Willi syndrome d) Cri-du-chat syndrome

54. In which phase of meiosis does genetic recombination occur? a) Prophase I b) Metaphase I c) Anaphase I d) Telophase I

55. Which of the following enzymes is responsible for repairing DNA damage and maintaining genomic stability? a) DNA polymerase b) DNA helicase c) DNA ligase d) DNA repair enzymes

56. What is the term used to describe a change in the DNA sequence that does not result in a change in the amino acid sequence of a protein? a) Missense mutation b) Silent mutation c) Nonsense mutation d) Frameshift mutation

57. In which part of the DNA molecule do genetic mutations primarily occur? a) Exons b) Introns c) Promoter regions d) Repetitive DNA sequences

58. What is the purpose of RNA polymerase in gene expression? a) To replicate DNA b) To cut DNA at specific sites c) To transcribe DNA into RNA d) To translate RNA into protein

59. Which of the following is an example of a post-transcriptional modification that occurs in mRNA processing? a) Splicing b) Replication c) Translation d) Translocation

60. What is the role of the ribosome in protein synthesis? a) To transcribe DNA into RNA b) To translate RNA into protein c) To replicate DNA d) To cut DNA at specific sites

61. Which of the following is a technique used to analyze gene expression at a single-cell level? a) RNA sequencing b) Microarray analysis c) Flow cytometry d) Polymerase chain reaction

62. What is the purpose of CRISPR interference (CRISPRi) in gene regulation? a) To amplify DNA fragments b) To cut DNA at specific sites c) To insert genes into cells d) To suppress gene expression

63. Which of the following is a method used to deliver genetic material into plant cells without the use of Agrobacterium? a) Biolistics b) Protoplast fusion c) RNA interference d) DNA microinjection

64. What is the primary source of genetic variation in plants? a) Mutation b) Recombination c) Natural selection d) Genetic drift

65. What is the purpose of transposons in plant genomes? a) To regulate gene expression b) To repair DNA damage c) To induce genetic mutations d) To move and insert themselves into different genomic locations

66. Which of the following techniques is used to study the function of a specific gene by suppressing its expression? a) RNA interference b) DNA sequencing c) Microarray analysis d) Polymerase chain reaction

67. Which of the following is an example of a plant hormone involved in plant tropisms, such as phototropism and gravitropism? a) Abscisic acid b) Gibberellin c) Ethylene d) Auxin

68. What is the purpose of DNA barcoding in plant identification? a) To amplify DNA fragments b) To cut DNA at specific sites c) To compare and identify plant species based on DNA sequences d) To insert genes into cells

69. Which of the following is an example of a plant defense mechanism against pathogens? a) Hormone production b) Cell wall reinforcement c) Production of antimicrobial compounds d) All of the above

70. What is the purpose of proteomics in plant biology? a) To study gene expression patterns b) To create genetic mutations c) To analyze and identify proteins in plant cells d) To control gene expression

ANSWERS

1. b) Plant genetics
2. b) Organism level
3. d) Genetic inheritance
4. a) Red flower color
5. a) Dominant phenotype
6. a) Molecular genetics
7. a) James Watson and Francis Crick
8. c) Nucleotides
9. b) S phase
10. c) DNA helicase
11. a) To amplify DNA fragments
12. c) Genetic engineering
13. b) To cut DNA at specific sites
14. c) A wheat plant with a disease-resistant gene inserted
15. c) Both mutations and sexual reproduction
16. c) Substitution of one base pair for another
17. c) Deletion of a chromosome segment
18. c) Gene expression
19. c) To start transcription of a gene
20. d) All of the above
21. b) It carries amino acids to the ribosome for protein synthesis.
22. b) DNA sequencing
23. c) To separate DNA fragments based on size
24. d) All of the above
25. b) To insert genes into plant cells
26. a) Agrobacterium-mediated transformation
27. a) A plant with genes from a different plant species
28. c) To reduce the use of chemical herbicides
29. c) Microarray analysis

30. b) To create genetic mutations

31. d) AFLP (Amplified Fragment Length Polymorphism)

32. a) To regulate gene expression

33. d) Gene flow

34. a) DNA sequencing

35. a) Hybridization

36. d) To regulate gene expression

37. a) To store and preserve plant genetic material

38. a) Loss of genetic diversity

39. c) Ethylene

40. d) To select plants with desired traits based on molecular markers

41. a) Bt cotton

42. b) To cut DNA at specific sites

43. c) Microarray analysis

44. c) To increase nutrient content in crops

45. b) Protoplast fusion

46. c) To insert genes into plant cells

47. d) Microsatellite marker

48. b) To cut DNA at specific sites

49. c) Rice with drought tolerance

50. b) To create genetic mutations

51. b) Pleiotropy

52. b) Polygenic inheritance

53. c) Prader-Willi syndrome

54. a) Prophase I

55. d) DNA repair enzymes

56. b) Silent mutation

57. a) Exons

58. c) To transcribe DNA into RNA

59. a) Splicing

60. b) To translate RNA into protein

61. c) Flow cytometry

62. d) To suppress gene expression

63. a) Biolistics

64. a) Mutation

65. d) To move and insert themselves into different genomic locations

66. a) RNA interference

67. d) Auxin

68. c) To compare and identify plant species based on DNA sequences

69. d) All of the above

70. c) To analyze and identify proteins in plant cells

PLANT BREEDING AND IMPROVEMENT

1. Which of the following techniques involves crossing two genetically distinct plants to create offspring with desired traits?
a) Plant tissue culture
b) Genetic engineering
c) Plant hybridization
d) DNA sequencing

2. What is the primary goal of plant breeding?
a) To increase genetic diversity in plant populations
b) To create genetically modified organisms
c) To develop plants with improved traits
d) To study gene expression patterns

3. Which of the following is a method used in plant hybridization? a) Self-pollination b) Cross-pollination c) Both self-pollination and cross-pollination d) None of the above

4. Which of the following is an example of a hybrid crop? a) Wheat b) Tomato c) Rice d) Corn

5. What is the purpose of backcrossing in plant breeding?
a) To develop plants with increased genetic diversity
b) To create genetically modified organisms
c) To introduce disease resistance into a plant variety
d) To study gene expression patterns

6. What is the term used to describe a cross between two individuals that differ in two traits? a) Monohybrid cross b) Dihybrid cross c) Test cross d) Self-cross

7. Which of the following is an example of a quantitative trait in plants? a) Flower color b) Leaf shape c) Plant height d) Disease resistance

8. Which of the following is a limitation of traditional plant breeding methods?
a) Lack of genetic diversity
b) Slow and time-consuming process
c) Inability to introduce specific traits
d) All of the above

9. What is the purpose of marker-assisted selection in plant breeding?
a) To create genetic mutations
b) To introduce genes into plant cells

c) To select plants with desired traits based on molecular markers
d) To control gene expression

10. Which of the following is a method used to induce genetic variation in plants? a) Mutation breeding b) Genetic engineering c) Tissue culture d) All of the above

11. Which of the following is an example of a genetic modification technique used in plant breeding?
a) Protoplast fusion
b) Hybridization
c) Marker-assisted selection
d) Backcrossing

12. Which of the following is a technique used to produce haploid plants for plant breeding purposes?
a) Doubled haploid production
b) Polyploidization
c) Transposon tagging
d) Micropropagation

13. What is the purpose of embryo rescue in plant breeding?
a) To replicate DNA
b) To cut DNA at specific sites
c) To rescue developing embryos from nonviable crosses
d) To amplify DNA fragments

14. Which of the following is a method used to improve the nutritional content of crops? a) Biofortification b) Mutation breeding c) Embryo rescue d) Protoplast fusion

15. What is the term used to describe a plant variety that is genetically uniform and stable? a) Hybrid b) Genotype c) Heterozygote d) Pure line

16. Which of the following is an example of an abiotic stress that can impact plant growth and productivity? a) Insect infestation b) Fungal infection c) Drought d) Genetic mutation

17. What is the purpose of gene banks in plant breeding?
a) To produce genetically modified organisms
b) To analyze gene expression patterns
c) To store and preserve plant genetic material
d) To control gene expression

18. Which of the following is an example of a traditional plant breeding method that does not involve genetic modification?
a) Mutation breeding
b) Hybridization
c) RNA interference
d) CRISPR-Cas9

19. Which of the following is a technique used to analyze and compare DNA sequences for genetic variation studies?
a) DNA replication
b) DNA sequencing
c) DNA hybridization
d) DNA cloning

20. What is the purpose of gene flow in plant breeding?
a) To increase genetic diversity
b) To create genetically modified organisms
c) To introduce specific traits into plant populations
d) To study gene expression patterns

21. What is the term used to describe the phenomenon in which the phenotype of a hybrid is superior to both parents? a) Genetic drift b) Inbreeding depression c) Heterosis d) Genetic recombination

22. Which of the following is a method used to determine the genetic similarity between different plant varieties?
a) RNA sequencing
b) Microarray analysis
c) AFLP (Amplified Fragment Length Polymorphism)
d) Gel electrophoresis

23. What is the purpose of gene editing techniques, such as CRISPR-Cas9, in plant breeding?
a) To increase genetic diversity
b) To create genetic mutations
c) To introduce specific traits into plant varieties
d) To control gene expression

24. Which of the following is an example of a self-pollinating plant? a) Corn b) Pea c) Tomato d) Apple

25. Which of the following is a method used to preserve plant germplasm for future use?
a) Gene banks b) Tissue culture c) Micropropagation d) Mutation breeding

26. What is the purpose of field trials in plant breeding?
a) To replicate DNA
b) To cut DNA at specific sites
c) To evaluate the performance of plant varieties under different conditions
d) To amplify DNA fragments

27. Which of the following is an example of a biotic stress that can impact plant growth and productivity? a) Herbivory b) Salinity c) Soil erosion d) Genetic mutation

28. Which of the following is an example of a seed treatment method used in plant breeding? a) Embryo rescue b) Micropropagation c) Seed priming d) Doubled haploid production

29. What is the purpose of polyploidization in plant breeding?
a) To create genetic mutations
b) To introduce genes into plant cells
c) To increase the number of chromosomes in a plant
d) To control gene expression

30. Which of the following is a method used to transfer specific genes from one plant to another? a) Gene flow b) Protoplast fusion c) Genetic engineering d) Tissue culture

31. What is the term used to describe the process of selecting plants with desired traits and allowing them to reproduce?
a) Natural selection
b) Artificial selection
c) Genetic drift
d) Genetic recombination

32. Which of the following is an example of a breeding method used to improve disease resistance in plants? a) Marker-assisted selection b) Embryo rescue c) Transposon tagging d) Backcrossing

33. What is the purpose of plant transformation in plant breeding?
a) To increase genetic diversity
b) To create genetic mutations
c) To introduce specific traits into plant varieties
d) To control gene expression

34. Which of the following is an example of a perennial plant? a) Corn b) Wheat c) Tomato d) Apple
35. What is the purpose of gene pyramiding in plant breeding?
a) To create genetic mutations
b) To introduce genes into plant cells
c) To combine multiple genes for improved trait expression
d) To control gene expression

36. Which of the following is a method used to improve crop productivity by crossing two different species or genera?
a) Hybridization
b) Inbreeding
c) Mutation breeding
d) Polyploidization

37. What is the purpose of end-use quality assessment in plant breeding?
a) To replicate DNA
b) To cut DNA at specific sites
c) To evaluate the suitability of plant varieties for specific applications
d) To amplify DNA fragments

38. Which of the following is a method used to enhance the nutritional content of crops through genetic modification?
a) Biofortification
b) Mutation breeding
c) Embryo rescue
d) Protoplast fusion

39. What is the term used to describe the removal of undesirable traits from a population over successive generations?
a) Selective breeding
b) Genetic modification
c) Genomic selection
d) Genetic purification

40. Which of the following is an example of a method used to improve drought tolerance in plants?
a) Marker-assisted selection
b) Embryo rescue
c) Transposon tagging
d) Genetic engineering

41. What is the purpose of progeny testing in plant breeding?
a) To create genetic mutations
b) To introduce genes into plant cells
c) To evaluate the performance of plant varieties through their offspring
d) To control gene expression

42. Which of the following is an example of a traditional plant breeding technique used to improve disease resistance?
a) Embryo rescue
b) Protoplast fusion
c) Mass selection
d) Gene editing

43. What is the purpose of gene expression analysis in plant breeding?
a) To replicate DNA
b) To cut DNA at specific sites
c) To understand how genes are regulated and expressed in different plant varieties
d) To amplify DNA fragments

44. Which of the following is an example of a breeding method used to increase crop yield?
a) Mutation breeding
b) Hybridization
c) Marker-assisted selection
d) Embryo rescue

45. What is the term used to describe the process of transferring genes from one plant species to another using Agrobacterium?
a) Agrobacterium-mediated transformation
b) Protoplast fusion
c) RNA interference
d) DNA microinjection

46. What is the purpose of gene introgression in plant breeding?
a) To create genetic mutations
b) To introduce genes into plant cells
c) To transfer specific genes from one plant to another through repeated backcrossing
d) To control gene expression

47. Which of the following is an example of a traditional plant breeding technique used to improve nutritional content?
a) Embryo rescue

b) Protoplast fusion
c) Mass selection
d) Mutation breeding

48. What is the purpose of cytogenetics in plant breeding?
a) To study gene expression patterns
b) To create genetic mutations
c) To analyze and understand the chromosomal composition of plant varieties
d) To control gene expression

49. Which of the following is an example of a breeding method used to improve crop adaptation to specific environments?
a) Marker-assisted selection
b) Embryo rescue
c) Transposon tagging
d) Recurrent selection

50. What is the term used to describe the process of creating a hybrid plant by crossing two different species or genera?
a) Interspecific hybridization
b) Intraspecific hybridization
c) Inbreeding
d) Self-pollination

ANSWERS

1. c) Plant hybridization

2. c) To develop plants with improved traits

3. c) Both self-pollination and cross-pollination

4. d) Corn

5. c) To introduce disease resistance into a plant variety

6. b) Dihybrid cross

7. c) Plant height

8. d) All of the above

9. c) To select plants with desired traits based on molecular markers

10. d) All of the above

11. a) Protoplast fusion

12. a) Doubled haploid production

13. c) To rescue developing embryos from nonviable crosses

14. a) Biofortification

15. d) Pure line

16. c) Drought

17. c) To store and preserve plant genetic material

18. b) Hybridization

19. b) DNA sequencing

20. a) To increase genetic diversity

21. c) Heterosis

22. c) AFLP (Amplified Fragment Length Polymorphism)

23. c) To introduce specific traits into plant varieties

24. b) Pea

25. a) Gene banks

26. c) To evaluate the performance of plant varieties under different conditions

27. a) Herbivory

28. c) Seed priming

29. c) To increase the number of chromosomes in a plant

30. c) Genetic engineering

31. b) Artificial selection

32. a) Marker-assisted selection

33. c) To introduce specific traits into plant varieties

34. d) Apple

35. c) To combine multiple genes for improved trait expression

36. a) Hybridization

37. c) To evaluate the suitability of plant varieties for specific applications

38. a) Biofortification

39. a) Selective breeding

40. d) Genetic engineering

41. c) To evaluate the performance of plant varieties through their offspring

42. c) Mass selection

43. c) To understand how genes are regulated and expressed in different plant varieties

44. b) Hybridization

45. a) Agrobacterium-mediated transformation

46. c) To transfer specific genes from one plant to another through repeated backcrossing

47. c) Mass selection

48. c) To analyze and understand the chromosomal composition of plant varieties

49. d) Recurrent selection

50. a) Interspecific hybridization

Made in the USA
Las Vegas, NV
18 October 2023

79298047R00144